The Institute of Biology
Studies in Biology No. 69

Genetics and Adaptation

E. B. Ford

F.R.S.

Emeritus Professor of Ecological Genetics
University of Oxford

Edward Arnold

© E. B. Ford, F.R.S., 1976

First published 1976
by Edward Arnold (Publishers) Limited
25 Hill Street, London W1X 8LL

Boards edition ISBN: 0 7131 2 2562 4
Paper edition ISBN: 0 7131 2 2563 2

Printed in Great Britain by
The Camelot Press Ltd, Southampton

General Preface to the Series

It is no longer possible for one textbook to cover the whole field of Biology and to remain sufficiently up to date. At the same time teachers and students at school, college or university need to keep abreast of recent trends and know where the most significant developments are taking place.

To meet the need for this progressive approach the Institute of Biology has for some years sponsored this series of booklets dealing with subjects specially selected by a panel of editors. The enthusiastic acceptance of the series by teachers and students at school, college and university shows the usefulness of the books in providing a clear and up to date coverage of topics, particularly in areas of research and changing views.

Among features of the series are the attention given to methods, the inclusion of a selected list of books for further reading and, wherever possible, suggestions for practical work.

Readers' comments will be welcomed by the author or the Education Officer of the Institute.

1976   The Institute of Biology,
                                         41 Queens Gate,
                                         London, SW7 5HU

Preface

It is obvious enough that a book of approximately sixty pages on a topic so immense as the Genetics of Adaptation must be selective to an extreme degree. There is no doubt in my mind what, then, its aim should be. It should provide information that is basic but not normally available to those for whom this Series caters.

We each have our own methods and preferences and, while by no means omitting illustrations, I have on the whole tended to sacrifice them to text for, with a fixed upper-limit, it is possible only to increase the one by reducing the other.

It is important to indicate what previous information is required in order to understand this book. It is that ordinarily gained by attending a few lectures on elementary genetics and evolution, or by reading one of the numerous introductions to those subjects that are on the market. That is to say; a knowledge of Mendel's two laws and their cytological basis, the facts of linkage and crossing-over, though not the detailed methods of calculating cross-over values, together with the concepts of mutation and of multifactorial inheritance. It is assumed also that the reader will know that genes have multiple, and often interacting, effects and that therefore dominance and recessiveness are conditions that evolve. On the evolutionary side, an elementary acquaintance with the idea of natural selection and of the importance of isolation is presumed. Yet there are a few topics essential for understanding the greater part of this book which are unfortunately omitted from most elementary courses of genetics. I supply information on them, sufficient for what is needed here, in a very short second chapter.

I have suggested various projects for amateur research. Some would be complete in themselves, others would not. For, as Major Leonard Darwin told me, his father, Charles Darwin, believed it to be a fine training to take even a transient part in a research of relatively long duration; such as can be maintained at a school or university.

In this matter, the first chapter stands apart from the rest. For it handles mutation-rate as an essential aspect of adaptation; and this is a subject that cannot appropriately be illustrated by photographs, nor is it suitable for practical work except at an advanced level. Yet it lays a foundation for a true perspective of genetics and in a way which I hope students may find stimulating though unusual. Furthermore, the value of relating the genetics of adaptation to Man is so evident that I have touched upon that matter in a brief final chapter. This also offers little scope for experiments at an elementary level: a loss for which its interest will, I believe, provide adequate compensation.

The scientist must aim to be a cultivated man, with a background of general education. The better he succeeds in this, the more percipient a scientist will he be, and I have allowed that thought to guide my presentation here.

I should like to express my gratitude to those who have taken up their time to give me an opinion on certain aspects of this book: to Professor P. M. Sheppard, F.R.S., Professor K. G. McWhirter and to Dr. E. R. Creed and Dr. E. R. Less, who also helped to correct the proofs.

I am indebted to Mr. J. S. Haywood for the photographs reproduced in this book. His skill as a photographer of natural history specimens, living or preserved, as of other subjects, is outstanding. I am fortunate to have had his collaboration here.

Professor G. C. Varley has been so kind as to loan from the Department of Entomology at Oxford the insects illustrated on Plates 4.1 and 5.4.

Oxford, 1975   E. B. F.

Contents

Preface   iii

1   Mutation-Rate   1

2   Polymorphism and the Super-gene   10
   2.1   Genetic polymorphism   10
   2.2   The super-gene   12

3   Pollution   14
   3.1   Melanism   14
   3.2   Genetic isolation and the flora of mine-tips   22

4   Isolation and Adaptation   25

5   Selection   37

6   Genetic Adaptations in Man   53

References   57

1 Mutation-Rate

The evolution of plants and animals is a product of natural selection acting upon genetic variability. Consequently, organic inheritance is basic to the evolutionary process and indeed from that point of view it must be judged. Yet it might well be thought that living organisms demand in this matter something incapable of attainment. For of necessity they must be endowed both with immense heritable *variability*, upon which selection can act, and great heritable *stability* to maintain the qualities, and combinations of qualities, that are of use to them. Here is a paradox that might seem impossible to resolve; yet resolved it has been, and by a set of adaptations so remarkable that they require our careful, though brief, consideration.

At the outset, we are confronted with the logical position that some sort of *units* responsible in some way for the qualities of the organism must be transferred from parent to offspring if heredity has any physical basis at all. Such units are today equated with the *genes*. Moreover, we have a simple mathematical proof obtained by means of correlation (FORD, 1973, p. 25), independently of any views on the nature of organic inheritance that we may hold, which demonstrates that the genes are contributed to each individual in approximate equality by its two parents (p. 8). At this stage of the analysis, we can shave away much that is superfluous with William of Ockham's razor: *Entia non sunt multiplicanda praeter necessitatem.* Evidently the simplest possible situation to secure equal bisexual inheritance is one in which the genes are present in pairs (the 'alleles'), whose members, being in no way committed to the male or the female line, are derived respectively from the pollen and the ovule, or the sperm and the egg.

We must at this juncture think of Mendel's two laws, imperfect though they are without the super-gene concept (pp. 12–13). Yet they demonstrate that heredity is 'particulate'. That is to say, the genes do not contaminate one another when brought into the same cell: and for this also, a universal proof is available, one that is applicable however many sets of alleles are involved, far beyond anything that can be demonstrated by simple Mendelian experiments: as in multifactorial inheritance (FORD, 1973, pp. 26–7).

It is now possible to turn back to the problem with which we started. How can organisms be endowed with the two opposed qualities of genetic variability and stability which are evident necessities of life, the very basis of adaptation? That is achieved by employing hereditary units which, on the one hand, are exceedingly stable and, on the other, can be recombined in an infinite variety of ways owing to segregation. One aspect of the permanence of the genes has already been indicated: they do not 'blend', retaining their identity when combined. But that is not enough; the large molecules, of deoxyribonucleic acid (DNA), which constitute them must be *intrinsically* stable, for the proposition that they do not contaminate one another would not ensure their permanence if they were 'unstable compounds' in the chemist's sense: to put the matter in genetic language, mutation must be rare.

Here we uncover one of the more famous of biological fallacies. The diversity between the different alleles at a locus is ultimately *supplied* by mutation, and selection can operate effectively upon the variability produced when they segregate; but those theories of evolution which are supposed to work by *controlling* mutation stand self-condemned. They include 'Buffonism', and 'Lamarckism', also, more overtly, the 'Mutation Theory' of de Vries. It is extraordinary that so much argument, and indeed experimentation, has been expended in an attempt to prove or disprove what is vaguely comprehended in the term 'Lamarckism'. Anything that comes to hand has been pressed into service to that end: the circumcision of Jews, the docking of horses tails or the effects of growing plants in differing environments. Such things are reiterated while the fallacy underlying the whole is left unexposed. That is to say, if mutations are predominantly responsible for heritable variability and the changes which bring about evolution, the genetic units are not stable enough to preserve the adaptations upon which the organism depends.

Moreover, 'Lamarckian' theories demand not only that mutations should be common but that they should be *directed*. This means that external agencies must influence both the soma and the germ cells: the latter in such a way as to mimic in subsequent generations and in a new environment, the effects produced on the body by the original environmental stimulus. A telling example illustrates the futility of that view.

When mice receive a heavy dose of penetrating radiation their hair follicles and associated pigment cells are damaged. Consequently, whether or not the hair falls out as an immediate result of treatment, on further growth it often becomes white. The one agent affects both the soma and the heredity of the animal for such radiation also causes mutation. Yet the result is not 'Lamarckian' (more correctly 'Buffonian') unless the rays were to single out always or to a disproportionate extent, the major gene responsible for coat colour and make it mutate to the allele giving rise to albinism. This they do not do; any gene, whatever its action, may be affected.

It is necessary during this discussion to keep in mind a few general properties of mutation. It affects single genes, or the occurrence of chromosome breakage and reconstructions, both in the reproductive cells and those of the body. Moreover it is recurrent, taking place more frequently at some loci than others, thus we can speak of the mutation rate at a locus. Also reverse mutations can occur, from the abnormal back to the normal allele, although in that direction the process is still rarer (p. 7); for in a complex system damage is easier than reconstruction. Yet the fact that the latter is possible indicates that the mutations we encounter are not essentially changes from a more to a less highly organized state.

It is important to use the correct terminology in this as in other aspects of genetics. *Mutation* is the (instantaneous) act of change in a gene or chromosome. The new unit which results from it must be referred to as a *mutant*; a term extended also to the organism if the change affects it. Thus the distinction between 'mutation' and 'mutant' corresponds to the one we make in logic between 'conception' and 'concept'.

Mutations are not related to the needs of the organism. Therefore they must nearly always be disadvantageous since it is very unlikely that a random change in a highly organized system, such as must be even a primitive plant or animal, will promote its harmonious working. Moreover, genes have multiple effects of the most diverse kinds. Consequently, if one component in the action of a mutant should chance to be favourable, it is indeed improbable that the others will be so too. Thus a 'favourable mutant' must be such that the advantage which accrues from one aspect of its working outweighs the disadvantages of the others. When that does happen, such a mutant will spread and be incorporated into the ordinary genotype, wholly or polymorphically (pp. 10–12); consequently its existence will then no longer depend on mutation. Also in view of the recurrent nature of mutation, an advantage must arise following a change in the genetics or the environment of the organism since the gene mutated in the past.

We are now in a position to evaluate natural mutation rate in the light of its adaptive significance. Here we meet the difficulties inherent in studying any very rare phenomenon, but the subject can in fact be approached from several different points of view.

Mutation in normal populations is known from its effects in the form of mutants. Some of these, if very disadvantageous, are wholly maintained by it, and on occasion one of them may be directly observable in very large experimental stocks. We have indeed no information on its average frequency per locus in the genotype of any organism; we may perhaps guess this to be in the neighbourhood of one in a million gametes; chromosome breakages and reconstructions are probably a good deal commoner. It must be remarked, however, that our direct knowledge of mutation is necessarily confined to the more frequently mutating genes, so that we inevitably much overestimate it.

Yet even at its upper limit, mutation is a very rare event. DOBZHANSKY (1970, p. 70) provides a table of comparative mutation-rates in a few organisms, including Man, in which sufficient information is available to give at least a correct idea of their order of magnitude, although the error involved in calculating them is great. The more rarely mutating loci are not of importance in the present context, while the estimates are too unreliable in respect of them; depending on one or two instances only, and fading off to those loci which mutate so rarely that the occurrence has never been observed. The upper limit of mutation rate should, however, have a meaning since it reflects the extent to which mutation can be reduced, if indeed that be possible; moreover, its assessment is reasonably accurate since it is based on larger numbers.

Setting aside a few exceptional genes (p. 6), we may evaluate mutation on the basis of its occurrence per million cells or gametes, and therefore per million individuals. The upper limit reached in each of the three multicellular species that have been most fully studied from this point of view provides a fair comparison, and it is a strange one. The most frequently mutating gene in Drosophila melanogaster is that for 'yellow body' estimated at 120 (per million); in the Maize plant, Zea mays the value is 106 for the I locus, having many effects one of which influences colour patterns; while in Man it is 190 for neurofibromatosis. (A condition in which tumours of the nerve sheath develop superficially and in the deep tissues.)

It seemed possible that relevant data on the House Mouse, Mus musculus, might have accumulated sufficiently for inclusion here. Dr. M. F. Lyon kindly informed me that the most up-to-date survey of spontaneous, as well as of induced, mutation in that species is one brought up to 1974 by Searle. It is clear that the results are approaching, but have not yet reached, the accuracy needed. The most mutable gene in M. musculus seems to be a (for non-agouti colouring), at 3 spontaneous occurrences in 67 395 gametes tested; the sexes combined. This is equal to 44.5 per million, but the error involved is so great as to make the estimate quite unusable. There appears to be a difference between the sexes in spontaneous (and induced) mutation-rate, which is slightly lower in the female. (The frequency for 7 loci combined in the male = 7.5×10^{-6} for 531 500 tested gametes. In the female it is similarly 6.1×10^{-6} for 164 999 gametes.)

It will be noticed that the maxima recorded in the three available species already mentioned (D. melanogaster, Z. mays and Man) fall within the same order of magnitude. But that fact at first sight appears impossible. For mutation is here judged per generation. Yet the period during which it can occur, and its effects assessed by the number of mutants produced, depends on the interval between the formation of the zygote and the reproduction of the individual to which that gives rise. Consequently, mutation should surely be estimated per unit of time; not per generation, which may indeed be long or short. Yet that simple and obviously reasonable conclusion proves unjustified. We may take the average lengths of a generation in the three forms under consideration as: Drosophila 14 days, Maize one year, Man 30 years. (Mean average age at birth of offspring, not of course age when the first child is born.) Judged per generation the mutation-rates are approximately similar in all three; judged per unit of time, they are wildly different. A mutation frequency of one in about 5300 individuals, being that for the neurofibromatosis gene in Man, when adjusted for time is equivalent to one in about 4 120 000 *Drosophila* generations. Judged per generation Man has much the same upper limit of mutation as the fly, while per unit of time the fly is nearly 800 times the more mutable. The Maize plant provides an intermediate step in this scale for, per generation, its mutability is of the same order as the other two, while in generation-length it falls reasonably well between them, somewhat nearer the *Drosophila* than the human value. As far as can be deduced, a similar situation obtains in all other multicellular organisms, plants and animals, in which an indication of it is available. Moreover, it will be noticed that the three reasonably well documented species discussed here, a higher plant, a fly and a mammal, constitute in classification a sample of living creatures so diverse that a quantity characterizing all of them must surely represent a great generalization. One thing only makes sense in this extraordinary situation: that mutation-rate is subject to selection, which pushes it back to a point when it becomes of negligible importance in *controlling* evolution; that is to say, to a sufficiently small value per generation, however long or short that may be.

Mutation is an obvious danger, since it takes place at random relative to the needs of the species and it is rare indeed for a random change to fit harmoniously into a highly organized system, as is that represented by even the simplest forms of life. Thus it is clear that mutation-rate is subject to counter-selection. How is this achieved?

Many genes are known to influence the general mutation rate (DARLINGTON and MATHER, 1949, page 155; IVES, 1950). Some of these are slight in their action while a few have large effects. They adjust the mutability of other genes to a varying degree and may give rise to chromosome abnormalities also. For instance, a high mutation-rate gene (*hi*) carried in the second chromosome of *Drosophila melanogaster* has repeatedly been obtained from wild stocks. It increases mutation in all the chromosomes. Its average overall effect when homozygous multiplies this ten times; when heterozygous, two to seven times. The variability of the latter value is due to the action of modifiers. Other genes are affected by *hi* to different degrees; 'folded' particularly so, and 'yellow' not at all. It gives rise to lethal and visible mutations in a ratio of 8 : 1. Moreover, the maximum inversion-rate in all chromosomes (if this can be taken as an index of their breakage-rate) is one in 400 germ cells when *hi* is homozygous, and one to four per 2000 germ cells when it is heterozygous.

Darlington and Mather (l.c.) point out that a gene of normal stability such as that for 'plum eye' in *Drosophila melanogaster*, may become highly mutable (so much so as to produce repeatedly a mosaic of mutant and non-mutant tissues) if moved near to the heterochromatin which, therefore, in some way enhances mutation. This is contained in a region, usually near the end of the chromosome, which gives evidence of possessing exceptional qualities; for it tends to condense early and stains more deeply with chromosome dyes. A few 'unstable' genes are known. These are excluded from the discussion of mutation rate on pages 4 and 5 because such loci seem to be very exceptional in animals and greatly restricted in plants, for the reason that they appear to be situated in or near heterochromatin. One of these is the R^r gene producing coloured aleurone in Maize. Another is that for the Dotted condition in the same species. This greatly increases the mutation-rate of a gene for anthocyanin production. Evidently heterochromatin provides a means for controlling mutation-rate in localized regions of the chromosomes. From these considerations one would deduce that some wild populations, both of species and races, are more mutable than others. And so we find in *Drosophila*, in which sufficiently accurate studies are available to estimate the point.

Though mutation is predominantly harmful (p. 3), genetic diversity is nevertheless ultimately supplied by it (p. 6). Thus one must envisage that its frequency is balanced between the needs so to reduce its occurrence that it cannot *control* evolution, while it can on rare occasions supply forms advantageous in special circumstances. The mutant in question will then be distributed through the population by selection. In rare instances one sees the need for this; once or twice for example in the melanism of Lepidoptera (p. 20).

Have we any evidence for such a balance in mutation-rate in addition to that already indicated? It seems possible, and in the following way. If its existence be a reality, then we might expect mutation to have been pushed back to a much lower frequency, compared with multicellular organisms, in forms which reproduce in immense quantities and very rapidly. There are some fairly definite suggestions of this, and here also we find it in highly diverse organisms.

The highest frequency of mutation in the bacterium *Diplococcus pneumoniae* is 0.01 per million, in *Escherichia coli* it is 0.006 and in *Salmonella typhimurium*, 0.005 per million; while in the mould *Neurospora crassa*, it is 0.03. It is to be noticed that these frequencies differ from one another by less than a factor of 10, though their highest value is about 10 000 times smaller than in the multicellular species already mentioned.

It has several times been asserted, never perhaps on unassailable evidence, that mutation-rate is greater in the offspring of a species-cross than in either of its parental types. That situation is certainly to be expected, since the genotype built up to check mutation will differ from one species to another. This is one of those matters in which genetic studies are lacking or imperfect and yet are especially needed.

KIMURA (1968) and others have argued that evolution is largely due to the chance survival of selectively neutral mutants, and it seems possible that they might wish to claim a higher mutation-rate for such 'neutral' genes. In the first place, there is no evidence for this. Secondly, it is not to be expected. The balance of advantage and disadvantage required to produce selective neutrality is exact (FISHER, 1930) and consequently it will not long be retained. It might, however, be thought that the 'neutral' genes in question are different in kind from others, affecting one character only, so that they do in fact long retain their neutrality. Again we have no evidence for such a thing. Moreover, as long as genes remain neutral as to advantage and disadvantage, they do not bring about adaptation and evolution.

Have we ever observed a mutation to a neutral or an advantageous gene? If so, is it unduly mutable? Even by 1968, as many as 992 genes had been studied in *Drosophila melanogaster*. Only one of them could possibly fall into the advantageous class. It is that for *ebony* body-colour, which sets up a balance with its normal allele (i.e. it is polymorphic, pp. 10–12) in experimental stocks. Dobzhansky (l.c.) gives its estimated mutation-rate as 20 per million. The error in these assessments must be great. But that for the highly unfavourable condition *eyeless* is 60 per million; and certainly *ebony* does not appear to have as high a mutation-rate as the disadvantageous *yellow body* (at 120 per million) already discussed.

It is interesting to notice that when GORDON (1935) liberated 36,000 heterozygotes for *ebony*, the most favourable mutant to have occurred in *D. melanogaster* stocks, it proved to be extremely disadvantageous in the wild. That fact suggests how selectively balanced the effects of a gene may be on approaching neutrality and how unreal it then is to talk of mutants as neutral or advantageous in effect.

The analysis given by FALCONER (1960), on his page 26, from the point of view of quantitative genetics, also supports the views developed here in this chapter. He points out that at the known spontaneous mutation-rates of organisms, mutation alone can produce only very slow changes in gene-frequencies. As he says, these might be important on an evolutionary time-scale but they could hardly be detected experimentally except with the use of micro-organisms. He also takes up the question of reverse mutation (from mutant to wild-type) and points out that this seems to be about one-tenth as frequent as that from wild-type to mutant. The equilibrium-frequency of such genes would therefore be about 0.1 of wild-type to 0.9 of mutant. That is to say, the mutant should be the common form and the wild-type the rare one. Since that is not so, gene-frequencies cannot be due to mutation alone. As Falconer goes on to point out, they are in fact due to selection.

There is one adjustment that may have to be made in considering the selective aspect of mutation-rate as we are doing here. As suggested to me by Dr. E. R. Creed, it is possible that mutation takes place chiefly at the replication of DNA (that is to say, the deoxyribonucleic acid of which the genes consist). If so, it should logically be assessed per number of cell-divisions. Accepting this for the moment, we should still evaluate mutation not per generation but per unit of time; though discontinuously, during the period from the formation of the zygote to reproduction. This would make a slight adjustment to what has so far been said.

However, there are several aspects to this matter. In the first place, its effect must be small in view of the relationship of mutation to length of generation, to which attention has been drawn. Secondly, Dr. Creed reasonably speculates whether a measure of mutation-rate per cell-division would greatly affect the male, which produces millions of sperm, compared with the female producing merely hundreds of eggs, or far less. Indeed the significantly greater mutation-rate in male compared with female mice might be an outcome of this, though the difference is very small (p. 4); and there is no indication of it in the comparison between *Drosophila*, giving rise to hundreds of eggs, and women with a smaller number and producing much fewer offspring.

It should be added that if one sex were contributing substantially to heredity compared with the other, by way of mutants to be removed by selection, one would not expect the equality of the genetic contribution of the two sexes that is demonstrated by correlation (p. 1). Also, it must be kept in mind that in the male we are in this matter dealing not with the millions of gametes formed but with the number of cell-divisions on the way to the production of any single gamete. Moreover, mutation might be checked in the male compared with the female line.

If we are to consider mutation-rate per cell-division compared with overall time, this might narrow somewhat the gap between the mutation-rate in bacteria and moulds on the one hand and multicellular organisms on the other. It should indeed be said that genetic research has not yet thrown any clear light on the contrasted possibilities mentioned here.

It might be thought that organisms could never become adapted to withstand the mutational effects of penetrating radiation. Yet this is what they certainly have done, a fact of which we have both direct and indirect evidence.

Significant genetic variation in resistance to the mutagenic action of radiation has been demonstrated both in higher and lower organisms: in mice (GRAHN, 1958), in *Escherichia coli* (GREENBERG, 1964) and in *Pseudomonas aeruginosa* (LEE and HOLLOWAY, 1965). Also more extensively in *Drosophila melanogaster*, in which WESTERMAN and PARSONS (1972) examined Co^{60} γ-rays in producing gene mutation and chromosome reconstructions in four inbred lines. These were found to differ markedly in such respects. PARSONS *et al.* (1969), using two dosages of Co^{60} γ-rays of, respectively, 90 000 or 110 000 rads, demonstrated such variation in wild populations also. They showed that the effects were largely additive and due mainly to genes in the second and third chromosomes, with a small contribution from some in the X.

Penetrating radiation is present in natural conditions owing to the existence of heavy elements in the rocks, the radioactive effects of which, varying much with local geology, are little known. It occurs also as cosmic rays, due to the radioactivity of the stars. Such rays include protons and neutrons. They are largely filtered out by the atmosphere so that their intensity increases with height; it is six times greater at 15 000 feet than at sea level. We may be sure that there will have been selection to diminish their mutagenic action at different altitudes. As CLARKE (1964, p. 267) wisely remarks, they must constitute a hazard in space travel, since they are highly effective in producing cancer through their impact upon somatic cells as well as on the germ-tract.

2 Polymorphism and the Super-Gene

The principles of polymorphism and of the super-gene, generally omitted from an elementary course of genetics should certainly be included in it, for without them our views of variation and evolution are grossly defective. Since those aspects of adaptation that are to be considered in this book cannot be understood without these basic concepts, they must briefly be explained in this short chapter.

2.1 Genetic polymorphism

This is a form of variation of great importance and universal application. It was defined by Ford, in 1940b, as 'the occurrence together in the same habitat of two or more discontinuous forms of a species in such proportions that the rarest of them cannot be maintained merely by recurrent mutation'.

The meaning and scope of that definition can best be understood by considering it piecemeal. Clearly it excludes both geographical and seasonal forms. It excludes also multifactorial variation responsible for some character which can take all values from one extreme to the other, both of which are rare while the mean between them forms a relatively common class: as in human height within any one population. There is, furthermore, another situation which cannot be polymorphic; it is that supplied by rare disadvantageous qualities eliminated by selection and maintained only by mutation. In mankind, a population is not polymorphic because it contains a few individuals who, like St. Mark, have 'stump fingers' (brachydactyly).

In polymorphism then, distinct alternative forms co-exist; as with the well known O and A blood groups which are not connected by intermediates and, in most races, occur together at a high frequency, both in men and women. Consequently, some sort of 'switch mechanism' must decide which out of two or more alternatives shall develop in any one individual. That control is nearly always supplied by the segregation of a pair of alleles, or by the alternative forms of a super-gene (pp. 12–13).

Relative to its effect, a mutant must be either advantageous, disadvantageous or of neutral survival-value. The celebrated mathematician Sir Ronald Fisher showed that we can exclude the latter situation as very exceptional. For he demonstrated that the neutrality of a gene requires for its maintenance a remarkably exact balance of advantage and disadvantage compared with its allele. He showed also that a neutral gene can only displace its allele at an exceedingly slow rate;

so slowly indeed that before it has advanced to any considerable degree, the delicate equipoise required for its neutrality will have been upset by genetic or environmental changes. Taking a step further, it may be said that a disadvantageous gene will be eliminated, its spread being checked at an early stage.

Thus we reach the conclusion that, except in a very small population, of a hundred or so (for the effects of chance are important only in very small populations (FORD, 1975)), any gene which has come to occupy, say, as much as 2% of a population must have some advantage. Why then does it not spread, producing indeed a *Transient Polymorphism* in the process, but one which disappears when it has reduced what was its normal allele to the status of a rare mutant?

To answer that question, we must first recall that genes have multiple effects and that mutations occur at random relative to the needs of the organism. Moreover, a random change is very unlikely to promote the harmonious working of a highly adjusted system such as the body of a plant or animal.

Consequently, on the rare occasions when one effect of a mutant does give rise to a useful character, its other effects will almost certainly be harmful. Thus, such a gene will spread to the point of neutrality, when its advantage is balanced by its disadvantages. This is most usually due to the fact that, in such circumstances, the heterozygote tends to gain an advantage over both homozygotes, so maintaining diversity and evoking a true *Balanced Polymorphism*.

We may well enquire how such a distinction between the genotypes is achieved. It is known that the advantageous effects of a gene become dominant and the disadvantageous ones tend to become recessive; therefore in the heterozygotes the advantages are operative and the disadvantages are not. Thus, relative to the gene in question, the two homozygotes will have some advantages and some disadvantages and are inferior to the heterozygotes which have advantages only. Other situations tending to promote heterozygous advantage also exist. (The accumulation of disadvantageous recessives near the locus of the gene controlling a polymorphism (FORD, 1975), and the existence of diversifying selection.)

From this brief analysis, two important consequences will be apparent. First, if a mutant having some advantage does arise, or if a gene gains some advantage and begins to spread, it is much more likely to become polymorphic than to displace its allele in the population. Therefore polymorphism is a common phenomenon; indeed it must certainly be operative if a gene occupies, say, 2% or more of available loci. Secondly, any polymorphism must inevitably be of importance: were it not, its controlling gene could not have reached even that low value. The latter conclusion is true even if its apparent effects are, or seem to be, negligible.

As an example of this, we may consider what appears to be a most trivial variation in Man: the (dominant) ability to taste low concentrations (150 parts per million) of phenyl-thio-urea. About three-quarters of the population in western Europe find this intensely bitter, while to one individual in four it is tasteless. But no one had the opportunity of sampling the substance until it was synthesized this century; yet the frequencies of the two types show the distinction to be polymorphic.

It has not yet been discovered on what grounds the gene recognized in this curious way is maintained in equilibrium by balanced selection. Yet we now know that it is associated with other and by no means negligible effects, for the tasters and non-tasters of this substance differ in the type of thyroid disease to which they are most liable (FORD, 1973). This at least makes it easier for us to appreciate that the gene in question does more important things than to promote the bizarre effect by which we detect it. Later in this book we shall, however, meet a number of instances in which the agencies controlling a polymorphism are clear and their significance evident.

In the light of what has just been said on balanced polymorphism, we are led to consider the meaning of an adaptive advantage when applied to that situation. We have seen that when a mutant or a rare gene gains an overall advantage and begins to spread, its frequency will increase in the population until it reaches the point at which it and its allele are balanced in the advantages and disadvantages which they confer: until, that is to say, they have become of neutral survival-value compared with one another. Such neutrality may be imposed ecologically, as in mimicry (pp. 45–9), or be genetic due to heterozygous advantage.

On pages 10–11 we were contrasting genes of neutral survival value with those that are advantageous. But is not that comparison invalid if a gene which is an asset can advance through the population to a point at which its value is obliterated, it and its allele being in equipoise? The answer is that there is a profound difference between the situation in which a gene of neutral-survival value is supposed to spread, and the neutrality that is maintained by powerful but contending selective advantages and disadvantages. For in the latter situation, either allele can rapidly adjust to a new frequency to meet changing conditions, whether genetic or environmental: a power absent from a gene which has not advanced through a population because of its relative importance.

2.2   The super-gene

A further aspect of polymorphism now claims our attention. For the unit controlling the alternative phases often takes the form of a super-gene, the nature and properties of which need briefly to be described.

The complex adaptations of an organism must often require for their control several pairs of alleles. When we consider the development of a structure fundamental to the body, the Mammalian heart, for instance, no special problem arises; for the alleles concerned in that process can each be maintained homozygous, since any rare disadvantageous mutants to which they are subject can be removed by selection. The situation is different when complex adaptations are polymorphic, for the appropriate phases of each must segregate together, as in butterfly mimicry (pp. 45–9). Something therefore must keep the correct genes of the different allelic pairs in association, and this is attained by close linkage. The smaller the cross-over value may be between two loci, the more surely does a given allele at one segregate with a given allele at the other. When the cross-over value between two loci is so low that they effectively act as a single unit, they are said to form a *super-gene*.

Occasionally, of course, crossing-over will separate genes which ought to co-operate, producing an ill-adjusted group which can, however, be eliminated in the same way as can a disadvantageous mutant. Genes exist which alter cross-over values by affecting the chiasma-frequencies along the chromosomes. However, the most effective way of holding a number of loci together is attained by enclosing them within an inversion. Crossing-over can take place unimpeded when neither or both the chromosome sections are inverted: when, that is to say, they are structural homozygotes. But when one chromatid in an homologous pair contains an inversion and the other homologous pair contains none (a structural heterozygote) an included cross-over destroys or impedes the tetrad. (Actually, the situation differs depending on whether or not the centrosome is included in the inversion; but structural heterozygotes tend to be eliminated in either event.) Consequently that type is eliminated.

Of course the genes needed to co-operate with one another may start on different chromosomes. They can then be brought on to the same one by occasional structural interchanges, transferring material from one chromosome to another. Such interchanges are not unduly rare. JACOBS *et al.* (1971) found 4 among the chromosomes of 2538 men; and probably the transference of minute fragments, generally concerned in super-gene formation, is considerably commoner.

The super-gene, then, is a device for holding together co-adapted groups of genes, and it must be considered an outstanding adaptation of organisms. We see it as one of the mechanisms for opposing Mendel's 'law' of independent assortment and we shall meet it repeatedly in subsequent chapters, in which its importance will be apparent.

3 Pollution

3.1 Melanism

Since the middle of last century, many moths in the industrial areas of Britain have become black. In some populations the process has advanced to the limit that can be achieved, in others it is in its early stages. The change in colour is sometimes multifactorial and is then slowly progressive. Yet in about a hundred species, exceedingly dark, often coal-black, forms have spread in and around manufacturing districts. Such 'melanism' is generally controlled by a single gene dominant in effect, though in a very few instances it is recessive.

This phenomenon constitutes one of the most striking, though not the most profound, evolutionary changes ever actually witnessed in nature; consequently it has been described many times (see especially, KETTLEWELL, 1973). There is, therefore, no need to do more than summarize its outstanding features here before mentioning some more recent results which throw further light upon certain aspects of it.

No butterflies are affected; and only those moths which rest fully exposed upon trees, fences or rocks, protected by their resemblance to bark or lichens. The enemies which they escape owing to their cryptic appearance are chiefly insect-eating birds which, as Kettlewell was the first to show, take a heavy toll of them; and, moreover, he proved that those which hunt by sight destroy their prey selectively.

The species which he principally studied is the Peppered Moth, *Biston betularia*. Its normal pale form is at a great advantage in unpolluted country, where lichens which it resembles to an astonishing degree abound and the vegetation is clean (Figs. 3–1, 3–2). Consequently, the intensely black phase, *carbonaria*, is particularly vulnerable there. The dark form survives better in or near manufacturing towns where the tree trunks are begrimed with soot but devoid of lichens, which are notably susceptible to pollution and cannot grow. Indeed the work of Kettlewell has demonstrated experimentally that those specimens which effectively match their background, be it light or dark, are the better concealed to avian as to human eyes.

Moreover there is some evidence that individual moths exercise choice in the matter. This does not mean that they fly to and settle upon appropriate dark or pale surfaces. Rather, when alighting upon a diversified background they shift about a little so as to match it the better, guided apparently by the contrast or accord between it and the hair-like scales on their head.

The experimental field-work on melanism was carried out first, and with outstanding success, by Kettlewell (l.c.). He bred large stocks of typical and melanic *Biston betularia* and liberated them on to tree trunks. Concealing himself in specially prepared hides, and using field glasses, he watched the fate of known numbers of pale and black specimens in rural and industrial areas respectively. By this means he demonstrated that those which do not match their background, whether light or dark, were subject to the greater predation by insect-eating birds such as Robins, *Erithacus rubecula*, Hedge Sparrows, *Prunella modularis*, and Redstarts, *Phoenicurus phoenicurus*. The destruction for which they were responsible proved to be severe, though its very existence had constantly been denied both by ornithologists and entomologists, so inept had been the work of many who had previously engaged in bird-watching and collecting Lepidoptera.

KETTLEWELL also marked large numbers of typical and of black male *B. betularia* with a dot of waterproof paint, placed on the underside of a wing so as normally to be invisible (females could not be used in this particular piece of work since they seldom visit a light). The insects were then liberated and a proportion of them was recaptured among samples obtained on subsequent nights at light traps; also in 'assembling traps', in which a virgin female constitutes the bait. The results indicated a heavily significant deficit of melanics in a rural area (Dorset) and of the pale form in an industrial one (the Birmingham neighbourhood). That is to say, the type whose colouring was appropriate to the locality survived the better.

CLARKE and SHEPPARD (1966) employed a different technique for this purpose. They killed, and preserved at −20° C., large numbers of the black and of the pale forms and attached them in lifelike positions on tree trunks by means of gum, leaving them for 24 hours. Their results confirmed those obtained by Kettlewell.

Similar observations could evidently be carried out by anyone interested in biology. Very few other species have been investigated by these means, so that there are great opportunities for amateur research here.

The results just described provide an obvious and apparently simple example of a genetically determined adaptation. Yet, as so often, a more penetrating examination of them shows their simplicity to be superficial; for breeding tests prove that the alternative colour-patterns of *B. betularia*, and indeed of many other species, are associated also with other and very different qualities. Not only may the pale moths grow faster than the dark during larval life, but the three genotypes of their controlling pair of alleles are not equally hardy. For as the melanic forms spread in industrial areas, initially on account of their appearance, the gene responsible for them becomes common enough for the heterozygotes to gain a physiological advantage (p. 11).

A specimen of the *carbonaria* form of *B. betularia* was first reported from

Fig. 3–1 *Biston betularia.* One of the typical form and one *carbonaria* resting on a lichen-covered tree: Dorset. (Natural size. With the kind permission of Dr. H. B. D. Kettlewell.)

Manchester in 1848. Yet by 1895, 98% of the population there was black. At that frequency its relative increase was halted; although since the pale phase is recessive, the heterozygous (black) specimens could then still occupy 24.3% of the total population. In that calculation we make two assumptions: that the three genotypes mate at random, which is nearly but not quite correct; and that they survive equally well, which is very far from the truth. We may consider the situation around Manchester when the melanics had become well established there. If we take the fitness of the (black) heterozygotes, Cc, as 1.00, that of the pale form was 0.50 while

Fig. 3–2 *Biston betualaria.* One of the typical form and one *carbonaria* resting on dark lichen-free bark: Birmingham district. (Natural size. With the kind permission of Dr. H. B. D. Kettlewell.)

for the black homozygote, CC, the value was 0.92 which represents its differential over the whole life-cycle. Yet the handicap of the homozygous dominant class is greatest in the perfect insect, in which the survival falls to 0.85. We see, then, why the spread of *carbonaria* has been checked in that district; a polymorphism has been generated in which the two colour phases are balanced at optimum frequencies relative to their viabilities as well as their appearance.

The major gene for wing and body colour must always have been mutating in the past, at its necessarily low frequency (pp. 2–6), to produce occasional black specimens: one is included in a collection of *Biston betularia* made before 1819. These would be eliminated by selection until the industrial conditions arose to which they were 'pre-adapted'.

Moreover, when the smoke of manufacturing districts has been reduced by legislation, the balance of advantage and disadvantage which maintains the polymorphism has readjusted itself. Thus CLARKE and SHEPPARD (1966) find that on the Wirral peninsula, Cheshire, the pale form, *cc*, of *B. betularia* was subject to a disadvantage of 41 to 55% prior to 1962. At that time a smokeless zone was introduced there, after which the handicap of the pale specimens declined to 21 to 23%; and a similar change is now occurring in Manchester.

It must not be supposed that the physiology associated with excess melanin production necessarily leads to heterozygous advantage. Black forms having poor viability in both genotypes have repeatedly been encountered.

Industrial melanism has become established in various parts of the world: for instance Japan, Germany and the U.S.A. In Britain its existence has been detected in many species though analysed in few, and we may well ask what components of the environment are responsible for its spread and maintenance. This matter has been examined by LEES *et al.* (1973) who undertook an analysis of melanism in relation to 14 environmental variables by means of multiple regression. The importance of crypsis in *B. betularia* has never been more clearly established in any insect than by the work of Kettlewell to which reference has just been made. The physiological asset of that species proves to be resistant to sulphur dioxide rather than smoke. For this reason, having become common in industrial areas in England, *carbonaria* has spread eastwards from them through Lincolnshire and East Anglia to the North Sea through apparently unpolluted country and still exceeds 50% on reaching the coast.

This is due partly to the fact that the sulphur dioxide content of pollution travels further than smoke. Moreover, it is probable that at its high frequency in industrial areas, the gene-complex has been able to evolve a satisfactory adjustment to this form of pollution so that it could spread more effectively in those rural conditions where a trace of the gas is to be found than when *carbonaria* first appeared.

A valuable comparison can here be made with three of the other species in which the occurrence of melanism has been studied. One of these is the Pale Brindled Beauty moth, *Phigalia pedaria*, in which the females are almost wingless though the males can fly and so disperse not the species but its genes. Yet melanism is detectable in both sexes and it has been the subject of detailed research by LEES (1971). He finds that there are three forms: a pale typical one and two melanics, less and more extreme, controlled by as many alleles; the typical is the bottom recessive and the more intense black is dominant to both the others. Lees shows that at their highest value the two melanic forms taken together amount to about 75% of the population, which they do in densely urban areas in England and South Wales. Yet in most rural districts their frequency is as low as 5%, occasionally less, though in some places it may be greater and can even reach 25%.

It seems in this, and in certain other insects, melanism has been established in advance of industrialization. Though in rural districts its value is no doubt derived chiefly from the physiological superiority of the heterozygotes, it may have some benefit owing to the behaviour of the predators of this moth. De Ruiter showed in 1952 that insectivorous birds tend to develop a 'searching image'. That is to say, they are inclined to hunt for specimens resembling the one they have recently found, even to the exclusion of others that are more obvious. Therefore, if common, palatable species obtain some advantage from mere diversity. That common moth the Green-brindled Crescent, *Allophyes oxyacanthae*, provides another good example of non-industrial melanism. It is widely dimorphic for cryptic and black forms both as an imago and a larva. It seems then that the presumably ancient melanism of *P pedaria* in rural areas is proving a further asset when industrial conditions are established. Smoke, not sulphur dioxide, increases the frequency of its melanism in manufacturing districts: a component demonstrating the value of its crypsis.

A comparable situation to that found in *Phigalia pedaria*, one which extends the analysis, has been successfully studied by CREED (1971) in the Coleoptera, using the Two-spot Lady Beetle, *Adalia bipunctata*. Here again are three common phases, of which the normal is recessive. It is scarlet with two black spots, one on each elytron. There are also two common melanics: black with six red spots, *sexpustulata*; the other black with four, *quadrimaculata*, which is the top dominant.

Here we have the remarkable situation of industrial melanism in a violently distasteful insect; one displaying warning coloration, which is probably always the more effective in the red phase. Thus the advantage of the black forms is derived purely from their physiology. As in *P. pedaria*, in which, however, crypsis is also important, that consists in superior survival relative to smoke pollution. In *Adalia* this operates only where the coal is volatile, and there the black specimens occupy up to 95% of the population. Where the locally mined coal is of low volatility, as in South Wales, melanism is usually rare as it is in rural areas. Creed finds some evidence for the view that the *quadrimaculata* form is about twice as common as the *sexpustulata* one where the local coal is strongly caking, as in the Manchester district and, to some extent, in Durham; but that these frequencies are reversed where weakly caking fuel is in use, as it is around Birmingham and in industrial Scotland.

It will be noticed that the frequencies of the melanic form of *Biston betularia* on the one hand, and of *Phigalia pedaria* and of *Adalia bipunctata* on the other, act as indicators respectively of the importance of sulphur dioxide and of the smoke components of pollution. This is a matter of at least potential practical importance (CREED, LEES and DUCKETT, 1973). It should be realized that these are merely the ingredients which have an outstanding effect in favouring the melanics; CREED *et al.* point out further environmental features that to a less extent contribute to this result.

It is notable that multiple alleles have become established at the major locus controlling melanism in all three of the species so far mentioned: a point to which only brief attention can be given here. In *B. betularia* at least three mutants producing less extreme melanism are known at the *carbonaria* locus. They all give rise to forms named *insularia*; which look like intermediates between *carbonaria*, to which they are recessive, and the typical insects, to which they are dominants. They appear intermediate also in the degree to which they are at a visual advantage and disadvantage respectively in unpolluted country and manufacturing districts. They seem well adapted to resting on boughs covered with *Pleurococcus* and so are favoured in the early stages of pollution, owing also to heterozygous advantage. On the other hand, they have never been known to exceed a phenotype frequency of 50% in England, for the homozygotes are subject to a heavy physiological handicap (see KETTLEWELL, 1973).

In *Phigalia pedaria* the less extreme mutant, known as 'intermediate' does not exceed a half, and rarely exceeds a third, of the melanics in any area. These intermediates must have a heterozygous advantage since their homozygote is of poor viability: that genotype may even be lethal. The respective advantages of the two chief forms of melanics in the beetle *Adalia bipunctata* have already been indicated.

In discussing briefly the theory of polymorphism it was pointed out that genes which interact to the advantage of the organism tend to become more closely linked, so as to form a super-gene. This holds certain alleles together or apart, as may be appropriate in distinct polymorphic forms. Kettlewell (l.c.) has provided important evidence which demonstrates the evolution of that condition.

The Oak Eggar, *Lasiocampa quercus*, is a large moth in which the males fly by day and the females only at dusk. It is subject both to industrial and non-industrial melanism, which is unusual in being recessive. On Rombold's moor, Yorkshire, which is heavily polluted by smoke from neighbouring manufacturing towns, about 2.4% of the imagines are very dark, though not black, due to the action of a major gene. Perhaps its spread results mainly from heterozygous advantage, though the perfect insects are much preyed upon by gulls, doubtless selectively. Melanic larvae are also present in this locality. There are two forms of them, both recessive: one in which the hairs are of a dark chocolate colour, instead of the normal reddish hue, and the other in which they are black. The latter is the rarer, but together they amount to 2% of the larval population, and Kettlewell shows that their relative frequency has increased since 1954.

Linkage has developed between all three of these genes. With independent assortment, no more than 2.4% of the melanic larvae, considered as a whole, should produce melanic imagines, yet 31.2% do so. At this frequency, about 1.7% of normal larvae ought to give rise to melanic moths, and that expectation is realized. Thus the evolution of a super-gene seems clear in this instance.

It is of much interest to notice that in the north-east corner of Scotland, where the heather is largely sparse and stunted and the general appearance of the moors is blackish owing to the exposed peaty soil, the imagines of this moth are also subject to apparently similar recessive melanism. It is far commoner there than in northern England, affecting about 70% of the population; having no doubt existed for a long time, while the insect is also subject to heavy predation by gulls. Kettlewell has crossed the Scottish and Yorkshire forms and finds that, though so much alike in appearance, they are controlled by different genes.

Larval melanism does not exist in north-east Scotland. Presumably it would be of advantage there but the required mutation upon which the effects of selection could operate has apparently not taken place, or has done so too rarely.

There seems evidence that in Yorkshire the species is subject to slight multifactorial darkening. This results in a general tendency towards a darker type, affecting melanics and non-melanics alike.

We find a similar situation in the moth *Gonodontis bidentata* in which the evolution of melanic polymorphism appears to have been held up in at least one district owing to the rarity of mutation. An intensely black form, *nigra*, occurs at a high frequency, over 75%, in and around Leeds and Manchester. It is established also in about 50% of the population at Cannock Chase, 20 miles north-west of Birmingham though in that city it is so far unknown; yet it would obviously be valuable since the imagines have undergone multifactorial darkening there. Indeed the existence of the *nigra* form of *G. bidentata*, though spread by selection, probably depends upon mutation in each area since the species has remarkably poor powers of dispersal (KETTLEWELL, 1973, p. 62). We have here an example of the situation mentioned in Chapter 1 in which it was suggested that evolution may occasionally be retarded by the necessarily low frequency of mutation. Indeed Kettlewell points out that a similar multifactorial trend is detectable in numerous moths of the family Caradrinidae.

Our knowledge of industrial melanism has been obtained principally by a combination of breeding work in the laboratory and ecological studies in the field: that is to say, by the technique of ecological genetics. The laboratory studies have involved not only ascertaining the straightforward genetics of the different forms but rearing segregating broods in an unsatisfactory environment, adjusted so as to produce a high mortality without destroying all the larvae. That procedure seems first to have been adopted by FORD (1940a) when investigating melanism in the moth *Cleora repandata*. Its object is to enhance any differences in viability between the genotypes. The method has since been used on a number of occasions, leading to the detection of significant departures from expected ratios, such as 1 : 1 or 3 : 1, in favour of the heterozygotes. An increase in the death-rate appropriate for this purpose can be obtained by semi-starvation.

The possibility of further useful work on this subject needs particularly to be stressed here. Melanism has seldom been recorded, and at all studied, in its initial stages. A great opportunity is to hand therefore if a dark or black form of a common moth begins for the first time to appear in any locality. Its frequency should be estimated at once, and subsequently in succeeding generations; when it may well prove to be increasing. We may envisage the situation also in which melanics start to appear in a district where they are isolated from a similar form either by an actual gap in the distribution of the species or by a region where only the pale insects are found.

It would be well worth while to breed melanics of any species from a region where they are common and back-cross them for several generations to specimens from a population in which they are unknown. The gene for melanism might then be found to have a detectably distinct effect in a gene-complex not adjusted to it. For example, the heterozygotes might then be distinguishable although dominance is normally complete. Extreme and highly informative examples of that result have been encountered in species-crosses in the course of two pieces of work on melanism; one, conducted by KETTLEWELL (l.c.) and the other by Cadbury (see FORD, 1975).

Random samples of the males of many moths may be obtained by means of a light trap. The specimens should be marked with a dot of waterproof paint when recorded. They can then be released, as they will be recognized and not counted again should they appear in the trap later. The material for breeding work would, in the first instance, have to be obtained by collecting larvae, for females rarely visit a light. Information procured by these means, whether in town or country, may provide opportunities for valuable research on adaptation and evolution.

3.2   Genetic isolation and the flora of mine-tips

It has in the previous Section been pointed out that, in addition to industrial melanism, melanic polymorphism may also occur in non-industrial areas, though at a much lower frequency. Consequently the opportunities for studying at least some aspects of the phenomenon are widespread. On the other hand, we now turn to a restricted occurrence which provides information of value on adaptation, including genetic isolation, although it may not often be available for investigation.

BRADSHAW, MCNEILLY and GREGORY (1965) have studied disused mine sites in North Wales where the spoil-tips are poisoned by salts of heavy metals: zinc, copper, tin and others. Even after forty years or more, the heaps of upcast are almost destitute of vegetation, looking like the landscape of the moon. Yet they found that such places are not absolutely bare. Here and there an occasional plant, especially of certain grasses, had established itself upon them, and Bradshaw and his colleagues wisely questioned how it had been able to do so. Laboratory tests showed that such exceptional individuals were tolerant of the particular metal contaminating the site while the same species growing in ordinary soil a few yards away was not.

The ability to withstand the poison proved to be genetic and multifactorial, though sometimes one of the genes concerned might have a more considerable effect than others; as in the grass *Festuca ovina*. The range of such variation is accordingly wide and allows a few slightly tolerant plants to exist in the normal population, and it will be from these that the resistant forms have been selected. Very distinct species, several grasses, *Plantago lanceolata, Silene inflata* and others, have been able to build up this type of adaptation, though some of those established even in the immediate neighbourhood have never been known to do so (*Dactylis glomerata*, for instance).

The conditions at Drws y Coed, a small copper mine (300 by 100 yards), have several times been quoted as providing a particularly clear picture of this type of adjustment. The shape of the valley in which it is situated tends to canalize the wind to a considerable degree from its prevailing direction, the west. Up-wind, plants of the grass *Agrostis tenuis* change from fully non-tolerant to fully copper-tolerant within one yard, at the edge of the mine. Down-wind, the reverse adjustment is achieved over 150 yards. We have here not only evidence of the effect of wind-blown pollen but of powerful selection in favour of the tolerant plants on contaminated soil and of their inferior performance in normal conditions.

Resistance to the various metallic ores is built up independently. SMITH and BRADSHAW (1970) point out that there is the possibility of reclaiming these unsightly areas by seeding them with appropriate tolerant grasses. They must also be provided with the nutrients, in the form of fertilizers which the sites lack.

In some instances, however, the poisoned soil has already become at least sparsely populated by genetically appropriate plants. A study of such populations throws an interesting light upon the ways in which adaptations may be built up.

ANTONOVICS and BRADSHAW (1970) examined the spoil-tip of a derelict lead and zinc mine on which the grass *Anthoxanthum odoratum* had been able to gain a footing. Tolerance to these metals began abruptly where the contaminated soil started. Yet culm-length changed gradually, as a cline (a 'cline' is a gradient between two adaptations; the term was introduced by Julian Huxley) between that characteristic of the adjusted and the normal plants, while flag-leaf width showed a 'reverse cline' as the boundary between the two types of soil was approached: that is to say, the distinctions between the two became more marked on nearing the line where they meet. This is a clear sign of powerful selection, which eliminates the less well adjusted intermediate forms that arise from crosses between those that are well adapted (p. 33).

Particularly relevant is the fact that the time of flowering is significantly earlier and the amount of self-fertilization much higher on than off the mine site. Here, then, we have evidence of incipient reproductive isolation: a subject to be discussed on pp. 25–8.

It would be well worth while to examine the spoil-tips of derelict mines in other parts of the country. It might be found that they are largely uncolonized by vegetation or, alternatively, that the species growing on them represent a small selection only of those established in the immediate neighbourhood. The soil on such sites should then be analysed to determine whether it contains poisonous substances. This could be done in the chemistry department of a school or neighbouring institution. Information on such soil analysis is given by GREGORY and BRADSHAW (1965).

If the ground on an old spoil-tip proves to be contaminated, plants from it and from outside the mine area should be grown experimentally on both types of soil, poisoned and normal. It would of course be very desirable also to test the genetics of resistance to deleterious ores if any indication of this were found. An extension along these lines of the important work carried out by Bradshaw and his colleagues in North Wales might well be a highly rewarding project. It may be added that adaptations of the kind which they detected have, on occasion, been important in the evolution of wild plants; such as those which survive the high magnesium content of serpentine rocks, as encountered on the Lizard peninsula of south Cornwall.

4 Isolation and Adaptation

It was outstandingly clear to Darwin, clearer even than the known facts justify (pp. 32–6), that isolation is essential for speciation and for many other types of evolution. In discussing this matter, it may at the outset be said that the chromosome mechanism can itself provide an efficient isolating system, and in two principal ways.

One of these is supplied by polyploidy. It will be recalled that in the ordinary diploid system every gamete is haploid; that is to say, it carries one chromosome of each type (the total being designated 'n'), while the additive nature of fertilization ensures that the body cells are diploid, possessing two such sets ($2n$). Occasionally, however, chromosome-reduction fails in meiosis, so forming a $2n$ gamete which, uniting with a normal one (n), gives rise to a triploid ($3n$). This is sterile because its chromosome pairing, which must take place between two alleles at each locus during the first meiosis, is unbalanced. If, however, two such $2n$ gametes unite, so as to form a tetraploid ($4n$) individual, fertility is largely restored because each chromatid then has a partner with which to pair. Yet it is not fully fertile at first owing to interlacing between the chromatids, though this can gradually be diminished in subsequent generations. Indeed, the attraction of homologous chromatids *in pairs* during the first meiosis ensures sterility to polyploids of uneven number (e.g. $3n$, $5n$) and at least a fair degree of fertility to those with an even number of chromosome sets (e.g. $4n$, $6n$). DARLINGTON (1963), in his brilliant discussion of the subject, points out that changes in ploidy are invariably upwards to greater values: so that organisms with haploid body cells arise only by parthenogenesis.

Polyploidy is of course far commoner in plants than in animals, since vegetative reproduction is normally required to establish a $3n$ population from which the tetraploid state may occasionally arise. As sexual reproduction is largely excluded in triploids, selection favours their vegetative propagation, which is indeed exceptionally prolific. This quality is so marked that it was at one time thought to be an automatic outcome of triploidy; it is, in fact, an adaptation.

Polyploidy, then, evidently creates a barrier to intercrossing. Consequently we sometimes find it established in a community adjusted to exceptional ecological conditions. This is particularly true when a plant reaches the extremity of its range in the sense that the environment has deteriorated relative to its needs: as may happen, for instance, in extending its distribution to the north or south. There, in situations reaching the intolerable, the only hope of survival lies in specialization, involving efficient adaptation to some particular set of conditions. Consequently polyploidy is of advantage to such races so that, cut off from crossing with the bulk of the species, they are free to perfect their adjustments owing to isolation. Evolution of that kind carries a population across the level at which it is judged specifically distinct; and for this there is justification for, of course, it is sterile with its parent form of lower ploidy since the 'hybrid' has an uneven chromosome set. Thus the Bird's Eye Primrose, *Primula farinosa*, ($2n=18$) passes into *P. scotica* ($2n=54$, being an 'adjusted' hexaploid) at the extreme north of its range in Britain: northern Scotland and Orkney. Its specific status has indeed become apparent since it has, not unnaturally, diverged somewhat in appearance; becoming dwarf, with flowers of a deeper pink.

The tendency for plants to attain polyploidy near the extremity of their distribution may lead to surprising results. DARLINGTON (1963) gives a striking instance of the kind. *Valeriana officinalis* ($2n=14$), a plant adapted to dry calcareous soils, exists as a diploid across the great plain of Europe. Near the north-western edge of its range it adapts by becoming a tetraploid and only that type has reached England, as a Holocene colonist after the last Ice Age and before the sea broke through at the Straits of Dover. Owing to the need for further adaptation, octoploids became established in some parts of the country. But these have the same chromosome-number as a related species *V. sambucifolia*, with which *V. officinalis* could accordingly then cross. Segregation among the hybrid offspring produced a vast array of forms, some of which have even been able to exploit damp situations and acid soils.

A second type of chromosome isolation is provided by heteroploidy, in which a chromosome is added to or lost from the normal set. This may arise from interchanges in which the greater part of one chromosome joins on to another leaving, in this instance, a centric fragment which, without the balance of enough chromosome material, gets lost. The result, of course, is a numerical reduction, so that the diploid number is odd. Alternatively, a chromosome may break through the centromere so as to create two centric parts. This again produces an odd total number, but by addition. An example of such 'unadjusted' heteroploids is supplied by Mongolian idiocy in Man, in which the chromosomes amount to a total of 47 instead of 46. Whether the result has been reached by gain or loss, there will be selection in favour of providing an even diploid number once more: this may be achieved by providing a partner for the compound one and eliminating the odd member, or by adding a partner for the single one if resulting from an addition. Either result can follow from a cross between two individuals each with an unpaired member. The result of such adjustments is shown in certain *Drosophila* species with chromosome complements in which $2n=6, 8, 10,$ or 12. This would seem to have involved reduction, since the higher values are present in the more primitive forms.

A suggestive instance of variation in chromosome-number is provided by the Mole rat, *Spalax ehrenbergi*, in Palestine (the true Mole, *Talpa*, an Insectivore, is absent from that country, in spite of Isaiah, 2.20). Here is a species in which adjusted local races reach the point of speciation (WAHRMAN *et al.*, 1969). Between Mount Hermon and the Sea of Galilee, and westwards to the Mediterranean from the line of the Jordan, the 2*n* number is 52. East of the Jordan in this northern region it is 54. From the Sea of Galilee southwards as far as Samaria, 2*n* = 58; while southwards again to the Negev, it is 60. The speciation is nearly complete, since though the territories occupied by these races are contiguous, the boundaries between them are clear-cut and few hybrids have been found. Thus the adaptive significance of this situation is strikingly suggested. It is probably related to the fact that the forms with the larger chromosome-complement occupy progressively more arid country.

That chromosomal situation, and others like it, is of much interest since it is open to experimentation, while it parallels one of the major steps in human evolution. In Man, 2*n* = 46, while in the three Anthropoids most closely related to him (the Gorillas, Chimpanzees and Orang-utans) it is 48. Yet again in the Gibbons, which are Man's next nearest relations, though a good deal further from him, 2*n* = 44. The comparison with *Spalax* is obvious.

We can now turn to those devices for achieving reproductive isolation that do not depend upon altering the chromosome numbers. One which may conveniently be considered at this point is the heterostyle-homostyle mechanism, which may again be illustrated from the Primulaceae. As DARWIN (1877) so fully described, the flowers of the Primrose, *Primula vulgaris*, and the Cowslip, *P. veris*, are normally of two kinds. These are pin (recessive) with a long style and anthers placed low down the corolla-tube, and thrum (heterozygous) with the positions of the male and female parts of the flower reversed. Evidently matings between the two give rise, as a series of back-crosses, to pins and thrums in approximate equality. As ordinarily described at an elementary stage in a botany course, it is easier for insects to bring about fertilization between the unlike than between similar flowers and, since pins and thrums are borne on different plants, that tendency promotes outbreeding. Yet it must be remembered that the fundamental distinction here is a physiological one, which leads to partial or complete sterility even when the cross pin × pin or thrum × thrum does take place: a widespread form of barrier often operating in the absence of any distinction in floral structure, the existence of which (that is to say, of heterostyly) tends merely to reduce wasteful crossings. Yet that additional morphological attribute is valuable enough to have evolved in eighteen orders of Angiosperms.

The genetic control of heterostyly is in fact a super-gene of at least seven loci. They include those for the pin qualities of long-style (*g*), pin stigma incompatibility (*i^s*), pin pollen incompatibility (*i^p*) and anthers placed low (a). Alternatively, those for thrum comprise genes for short-style (G), thrum stigma incompatibility (I^s), thrum pollen incompatibility (I^p) and anthers placed high (A). Thus the two genotypes include the genes: g, i^so, i^p, a for pin and G, I^s, I^p, A for thrum. One of the two cross-overs at the appropriate place within the super-gene associates g, i^s, I^p, A produces a long homostyle (the reciprocal cross G, I^s, i^p, a, gives rise to the short homostyle, with stigma and anthers down near the bottom of the corolla tube). That is to say, a self-fertile flower which has a long style clasped by high anthers both at the top of the corolla tube. The latter arrangement avoids the transference of pollen by insects, while the incompatibility genes in regard to stigma and pollen are dissociated so that the flowers are self-fertile.

Since the homostyle type (whether long or short) is largely self-fertilizing, it tends to maintain adjustments to local conditions whether at the edge of a species range or in exceptional habitats. The most efficient evolutionary plan therefore would be for a marginal plant-community first to become polyploid. This allows variation within it, on which selection can act without contamination from the normal form 'behind' it, as it were. When appropriate adaptations are achieved, they could be preserved in a heterostyled species by taking the further step to homostyly. This appears to have occurred in *Primula scotica*, already mentioned; for it is not only a polyploid but the only member of the British Primulaceae which is always homostyled; which its progenitor *P. farinosa* is not, being distylic.

Organisms must normally evolve and adapt to fluctuating local conditions by means of selection operating on genetic variability. If, however, they become particularly well adjusted, almost all their variation must be disadvantageous; and there will then be strong selection for uniformity and for the means of maintaining such. Yet if that state be irreversible, they have sacrified long-term evolution for short-term advantage, since they can no longer conform to changing ecology and must ultimately perish. The exceptional asset of the heterostyle-homostyle arrangement is here evident, for it allows a plant species to pass back and forwards, from relative variability to relative stability, at need. It is therefore a particularly favourable evolutionary device.

Thus in Britain the primrose, *P. vulgaris*, is normally a distylic plant. Yet there are at least two populations containing a large proportion of (long) homostyles. One of these is on the Chilterns. The other, which is the better known and more easily worked, because the species is far commoner there, is near Sparkford in Somerset. Each is thought to occupy an area of about 11 by 12 miles. It is not yet clear what ecological conditions cause this trend towards uniformity in these two districts. However, the English countryside has departed so far from its primeval state, and is subject to so much local diversity, that it is likely to include some areas where it pays the primrose to readjust its reproductive system.

Presumably it is particularly well adapted to the local ecology in these two localities.

Flowers which superficially resemble short, but not long, homostyles are not infrequent among normal primroses. These are generally pins with a crumpled style which brings the stigma down to the low anthers, but they have not, of course, the genetic arrangements to produce self-fertility. CROSBY (1940), who discovered the two populations containing a high proportion of homostyles, maintained that no true examples of this condition exist as rarities in the ordinary British population of the species, but we now know that in this he was incorrect (FORD, 1975). It will be from such occasional rare cross-overs that the true selfing-homostyles can be supplied at need.

The heterostyle-homostyle mechanism of the primrose is described in detail, with a diagram and photographs, by FORD (1975, Chapter 10). It would be a fascinating study to look for additional areas where the homostyle form has spread and, if discovered, to analyse the frequencies of the floral types there. Also, similar investigations are much needed in the cowslip, in which such populations have not yet been found though it is highly probable that they exist.

Since isolation by polyploidy can seldom occur in animals and nothing akin to the heterostyle-homostyle system exists in them, their methods of exploiting extreme conditions are more wasteful, though widely used by plants also. That is to say, those which begin to develop local adaptations may attain some success and those that do not may fail. Therefore the group that achieves initial adjustments is often left to perfect them in isolation, shut off geographically from the rest of the population (but see p. 33).

We constantly find therefore that animals as well as plants form local races at the edge of their range, and these may have quite abnormal characteristics. Two out of innumerable instances will in the first place serve to illustrate the point.

The moth *Malacosoma castrensis* can as a caterpillar feed on many common plants and the species is widespread in a variety of habitats in western Europe. Yet in England, where it is evidently a Holocene colonist, it is confined to salt marshes along the Thames estuary and up the coasts of Essex and Suffolk, surviving by adaptation to that one highly specialized environment. There it has invented a new means of dispersal suited to its peculiar needs. The females, though fully winged rarely, if ever, fly. Their eggs have become resistant to brine and are laid in batches on debris at the water's edge. This is washed round the salterns at spring tides and may be left above the normal high-water mark. On hatching, the young larvae crawl to nearby food and may thus establish themselves in a new area. Yet in Continental Europe this moth is not at all restricted to the coast nor to marshes, maritime or otherwise and occurs in central France and Germany.

It is obvious that in any population adjusted to a specialized environment selection will oppose the scatter of individuals even in a species with powerful means of dispersal. For example, the Swallow-tail butterfly, *Papilio machaon*, is a large strongly-flying insect widespread in western Europe. It is often seen ranging over the countryside there, and to a considerable altitude in mountainous regions. Its distribution in Scandinavia extends to the North Cape, but in England it reaches its north-western limit. Here, as a permanent resident, it is confined to the 'broads' and fens of Norfolk, and until the 1939–45 war to a small similar area at Wicken near Cambridge. In this country it has evolved a race, *britannicus*, differing in appearance slightly but characteristically from that found in France. Moreover, the larvae feed only on the Marsh Milk Parsley, *Peucedanum palustre*, to which they are not at all confined elsewhere, generally eating wild and cultivated carrot.

A number of different races of the butterfly occur in Europe, but the British has acquired distinct habits. For this confines itself entirely to its specialized environment, though flying powerfully within it, and does not wander into the neighbouring countryside. Selection has thus produced an adaptation in behaviour essential to a species with sharply localized requirements (FORD, 1975).

It is indeed common to encounter genetically determined differences in habit; several instances of the kind are mentioned in this book. Thus the material upon which selection can operate to adjust the reactions of organisms to the stresses of their environment lies ready to hand.

In neither of the two instances just mentioned have steps been taken to detect any of the genetic differences that must have accumulated between the British and Continental races, or if they have actually begun to diverge specifically. We may, however, look at two further examples in which crosses between geographically isolated populations have fortunately been studied.

The Muslin Moth, *Cycnia mendica* (Arctiidae), is common in the south of England but rare in the north and in Scotland. It is widespread in Ireland. In Britain the sexes are markedly dimorphic, for the males are dark brown and the females white; while both have a sparse scatter of black dots over the wings. The British and Irish females are not distinguishable, but in the Irish race the males resemble them in colour, being white (Fig. 4–1). This is the form *rustica*. It occurs also in a few places in Continental Europe. Crosses between English and Irish specimens show that the difference between the chocolate-brown and the white males is due to a single gene. Neither colour has gained dominance, evidently because isolation has prevented the formation of heterozygotes on which the necessary selection could act. These can of course be obtained experimentally. They prove to be highly variable, the mean being of an intermediate sandy colour. There may well be racial differences also between the apparently similar females, perhaps influencing protein

Fig. 4–1 *Cycnia mendica*. Numbered in two horizontal rows. (a), English male. (b), English female, (c), *rustica*, being the Irish male (the English and Irish females are indistinguishable). (d), Heterozygous male from a cross between the English and Irish races. (Enlarged: actual size of a, from wing tip to wing tip, = 3.2 cm).

variation; if so, this might be detected in them, and in the males also, by electrophoresis (pp. 36, 51). Had the females the racial colour-distinction of the males, it would almost certainly have been easy to select brown to become dominant in one line and white in another. As it is, that result would be slow to attain artificially since the potentiality of the females cannot be identified visually.

There is an interesting feature in the ecology of this moth. In England, the conspicuous white females often fly in daylight, almost certainly protected by their resemblance to the highly distasteful White Ermine, *Spilosoma lubricipeda*, which constantly sits in conspicuous positions during the day (see the section on Mimicry). To this the males bear no resemblance and accordingly they are on the wing only from dusk onwards, when their dark colouring makes them inconspicuous. We do not know if the white *rustica* males accord with the day-flying habit of the females, which their appearance would justify. It is to be hoped that observations in Ireland will establish the point as, in either event, it would throw light upon the evolution of habit.

A species may be broken up by its ecological requirements into isolated groups distinct enough to promote their separate evolution. The Grass Eggar, *Lasiocampa trifolii*, a large moth in which the males fly by day and the females at dusk, provides an example of this kind, one which has been studied genetically. The species is found near the sea and is subdivided into a number of populations along the south coast of England. In

Cornwall the imagines are dark reddish-brown while at Dungeness, Kent, they are of a pale putty colour. Dr. H. B. D. Kettlewell (personal communication) showed that this is a unifactorial character in which the dark shade is dominant. He also discovered that the males of the F1 generation are normal but the females intersexual. That is to say, these two races have taken a definite step towards speciation. Thus they illustrate a rule due to J. B. S. Haldane: that if in a cross between two populations or species one sex is abnormal, rare or absent, it will be that with the XY chromosome-pair. The reason for this is not difficult to follow. It is a curious fact that though the aims of sex-determination are similar in most animals, the production of clearly distinct males and females in approximate equality, yet that result may be achieved very differently in one group, or even species, from another. However, the basic plan seems to involve the presence of a large number of genes, scattered among the chromosomes, which tend to promote respectively either maleness or femaleness. The total effect of these is contrary in the autosomes and X-chromosomes. Furthermore, the Y-chromosome is normally inert as a sex-determining agent, though it is not so in Man.

The action of such genes can, and often does, vary quantitatively from one animal species to another, for they work effectively so long as the combined sex-control of the autosomes outweighs that of one X-chromosome but is outweighed by that of two. On crossing two races, a mean value is taken by the sex-determining genes, both in the autosomes and the X-chromosomes, in the homogametic sex (XX), in which one member of each chromosome-pair is derived respectively from the two parents. That, however, is not so in the heterogametic sex (XY), which possesses an X-chromosome from one parent (and race) and none from the other; consequently the sex-control is here thrown out of balance. In many animals, for instance the Mammals, the XY pair is carried by the male; but in butterflies and moths, and in birds, by the female. Therefore it is the latter sex which gives the first signal of incipient speciation in the Lepidoptera.

We have so far been thinking of adaptations arising in isolation, whether chromosomal or geographical; but certain populations can also be adjusted to differing habitats within their range by means of a character-gradient or *cline* (p. 24). Its adaptive significance in any one instance may not be obvious. For example, the sedentary fish *Zoarces viviparus* is found in Norwegian fjords and, for no apparent reason, its vertebrae are steadily reduced in number the further the animal lives from the open sea.

Often, however, the significance of a cline is evident. Thus the progressively greater size of many bird species at increasing altitudes, or in colder localities, is no doubt related to the maintenance of the high body-termperature. For heat-loss is of course reduced as surface-area becomes smaller relative to volume, as it does as an animal grows larger.

However, a continuous population may generate discontinuity, affecting sometimes one of its characters, sometimes many; whether or not these exhibit a cline on one or both sides of the boundary between them (p. 23). That situation can arise in several ways, and two of them should be mentioned here.

In the first place, when the races of a species are isolated, they may develop favourable adjustments to their own environment. Such successful populations will sometimes spread until they meet. Yet this does not necessarily mean that they will then merge to form an unbroken cline. On the contrary, when individuals carrying co-adapted but different gene-complexes interbreed, the hybrid between them will be ill-balanced and on further crossing will break down into an array of ill-adjusted forms. These, constantly produced at the interface between two races will as constantly tend to be eliminated by selection which, conversely, will preserve the satisfactory genetic structure of the contributing populations.

For instance, the Bordered-white Moth, *Bupalus piniarius* (Selidosemidae), is of the Scandinavian form in Scotland and northern England, having probably arrived there following the fourth Pleistocene glaciation. In the south of England the insect is of the more southerly European type which no doubt entered the country subsequently, as a Holocene colonist before the sea broke through at the Straits of Dover about 8000 B.C. The two races meet at the level of Shropshire and Cheshire. Though they hybridize there, with considerable variability, they have not merged but retain their identity except in a relatively narrow inter-breeding zone of 30 miles or so.

Secondly, a more subtle form of adaptive discontinuity is now being studied intensively. We need here to notice that, on the whole, variation may be either 'stenoplastic' adjusted, that is to say, to a restricted tolerance of environmental conditions, or 'euryplastic' in which it is adjusted to a wide one.

As clearly indicated by HANDFORD (1973), when an organism such as a butterfly occupies a large tract of country, many components of its environment may change from one part of its range to another; some sharply but others gradually and to a considerable extent independently. To these the organism needs to adjust itself. Yet it is very unlikely that the diverse aspects of the genetic variability which it must employ will act independently of one another too. For experimental genetics has taught us little indeed if not the lesson that genes have multiple effects and interact with one another to produce the qualities, of morphology, physiology or habit, for which they are responsible. Thus a species may be unable to follow in detail the ecological situations to which it is exposed. On the contrary, it may be forced to respond to them by passing abruptly from one to another of the somewhat limited series of balanced genetic systems which allow it to adjust to the average conditions it encounters over a considerable tract of country. There will come a position at which it must switch to another gene-complex, also of the euryplastic type, to adapt it to a further diversified region; the place at which it does so changing, perhaps, with differences in climate from year to year. Let us see how this conclusion accords with some ascertained facts.

The butterfly *Maniola jurtina* (Satyridae) has one generation annually, extending from late June to mid-August. It is widely distributed in Britain up to the north of Scotland at elevations below about 270 metres (800 feet).

Within the outer margin on the underside of its hind wings, small black spots may be seen. They may be absent, or present up to a total of five. We may here restrict ourselves to the females which are the more distinct from one region to another (the males have nearly always a single maximum at two spots (see Figs. 4–2 and 4–3). The spot-number is under multifactorial control, and its heritability lies within the range of 49 to 77% at 22°C. That is to say, taking the average value, 63% of the variation in spot-frequency is genetic and 37% environmental. Thus selection has a considerable component on which to work.

Across southern England, from the East Coast to mid-Dorset, female spotting remains the same, with a single maximum at no spots (Fig. 4–2). In east Cornwall it takes a different value, with two maxima: one at no spots and the other at two (Fig. 4–3). In each area spotting is stabilized in spite of great environmental diversity. This is particularly striking from mid-Dorset eastwards, where the butterfly may inhabit dry valleys on chalk and limestone Downs, marshes, open country on an acid soil and woodland rides; differences to which the spot-frequencies respond not at all. Yet one type of spotting may be replaced by the other, without any obvious relation to physical conditions or ecology. It may do so anywhere within the so-called 'boundary region', which extends from mid-Dorset to the Devon–Cornwall border. In some seasons we find the spectacular situation in which the change occurs within a few yards; in others, it takes place over a considerable tract of country, 20 miles or so, within which the spotting-types are intermingled in different localities.

Should the more abrupt of these transitions persist at the same place for some years, it may form a *reverse cline* (p. 24). That important situation is easily understood. We may quantify female spotting as the average number of spots on one hind wing in a given population. Evidently that average will be lower in eastern England, where the commonest condition is the spotless one, than in east Cornwall, where there is an excess at two spots as well as at none (0.95 in the south-east of England and 1.20 in east Cornwall are quite usual values). In a reverse cline, the distinction between two forms becomes greater as we approach the line where they meet. This demonstrates powerful selection, which favours, respectively, the two stabilized genetic types and tends to eliminate the intermediates between them, ill-balanced as these must be.

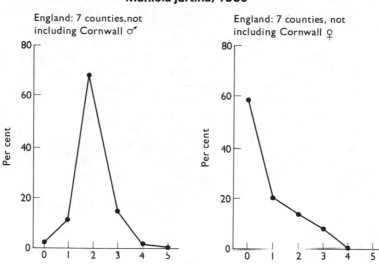

Maniola jurtina, 1950

England: 7 counties,not
including Cornwall ♂

England: 7 counties, not
including Cornwall ♀

Fig. 4–2 Spot-frequencies of *Maniola jurtina*; Dorset eastwards.

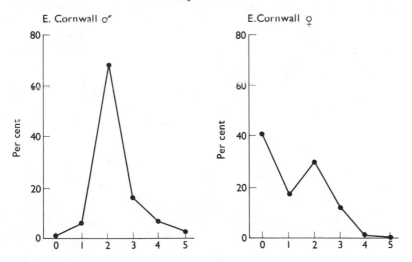

Maniola jurtina, 1950

E. Cornwall ♂

E.Cornwall ♀

Fig. 4–3 Spot-frequencies of Maniola jurtina; East Cornwall.
(Both from FORD, E. B. (1975). *Ecological Genetics*, 4th edn., with the kind permission
of Chapman and Hall, London.)

Looking back, we can see how clearly the situation in *Maniola jurtina* accords with the general proposition laid down on pp. 33–4. It seems that the 'boundary region' may be regarded as an area of instability in which the average genetic adjustments of the species east and west of it reach as it were a 'breaking point'. Consequently they are likely in that intermediate situation to become less securely entrenched and less resistant to annual differences between the seasons.

These facts obviously indicate the operation of powerful selection. Yet the criterion originally used to detect this, being the one so far considered, is trivial indeed: variation in the number of minute spots on the underside of the hind wings. Have we any reason to suppose that this is the outward and visible sign of important genetic adjustments? Fortunately such information is indeed available; for on studying other characters we find that some of them also break down and become unstable in this very 'boundary area'. That is true of the *position* of the spots as well as their number, and we have evidence that these two variables are independent (FORD, 1975). Furthermore, HANDFORD (1973) has detected two esterase polymorphisms in this insect; he did so by physico-chemical means, using the technique of electrophoresis. Yet these physiological variants, so profoundly distinct from spot-number, also show excess variability within the region from mid-Dorset to the Devon–Cornwall border.

Evidently the situation indicated here, in which a discontinuity arises within the continuous *M. jurtina* population in the south of England, is an example of a general proposition, and it may be reasonably enquired if we have evidence of it elsewhere within that insect's range. Indeed this is so. That same phenomenon has been encountered also on the island of Great Ganilly, Isles of Scilly, and in Perthshire. Therefore, as might be anticipated, it exemplifies one method by which organisms can adapt themselves to their environment.

5 Selection

It is a matter of much consequence to decide whether the effects of selection in adapting organisms to their environment can be studied in wild populations. Darwin thought they could not, except as a very long-term project: he envisaged fifty years in species reproducing annually. Owing to recent work on ecological genetics, we now know he was wrong. We may, therefore, first consider instances which demonstrate that such effects can certainly be both detected and measured; quite apart from discovering in what their advantages, or their balanced advantages and disadvantages, consist.

In 1930 Moreau, a well-known ornithologist, gave his opinion on the speed at which birds can evolve. Now the House Sparrow, *Passer domesticus*, was introduced into the U.S.A. from Europe in 1852. It has become abundant there and has spread widely over North America, adjusting itself to local conditions in various ways. By 1933 it reached Mexico City, where it has since evolved a distinct geographical race. Moreau had concluded that the minimum time required for a bird to achieve that evolutionary step was 5000 years. The time taken by the Sparrow to do so was 30 years. We can here judge the value of speculation, compared with observation, in analysing evolution.

We may now appropriately return to *Maniola jurtina*, the butterfly discussed at the end of the previous chapter. SCALI (1971) and his colleagues have discovered the way in which the imagines of this insect survive the heat of the Italian summer, for it is on the whole rather a northern species. Near sea level and up to an altitude of several hundred feet, it emerges from the pupa in late May and early June. Pairing takes place at once, when the males die after a life of a few days. The females then aestivate by secreting themselves in bushes, where they remain quiescent throughout late June, July and August. Thus the butterfly disappears during high summer. It becomes active again in early September as if a second hatching took place then, as was until recently believed. However, this flight, of course, consists almost entirely of females, which then fertilize their eggs with sperm received two-and-a-half or three months before and retained in their sperm-sac.

The spotting of *M. jurtina* on the underside of the hind wings has already been mentioned (p. 34). When evaluated before and after aestivation, it appears that a disproportionately large number of the higher spotted specimens fail to survive the hot weather. For on emergence from the pupa the population has a spot-maximum at 1, with high values at 0 and 2, giving a spot-average of 1.35. After aestivation, there is a large excess with no spots, and the spot-average is then 0.91: indeed the spot-frequency at that time much resembles that shown in Fig. 4–2. From these data Scali (l.c.) has found that an average selection of 64% operates against females with 2 to 5 spots during their period of inactivity.

Since this elimination occurs each year as a regular feature of the life-cycle, it must be balanced by a correspondingly better survival during the larval and pupal stages of those individuals destined to become high-spotted; otherwise it would not be long before these were eliminated from the community. This is an example of *endocyclic selection*, in which genes pass from being advantageous to disadvantageous, and the reverse, during the life of each individual. It is to be compared with *cyclic selection*, in which the advantage and disadvantage of certain genes is, for example, reversed during spring broods compared with summer and autumnal ones; a situation demonstrated, for instance, by BIRCH (1955) in the fly *Drosophila pseudoobscura* in California.

The phenomenon of aestivation in *M. jurtina* does not occur north of Italy, where adjustment to weather conditions is secured by an extended period of emergence, lasting from late June until mid-August. That device is also employed by the Italian population at high altitudes of about 700 metres (2000 feet) and upwards, while at intermediate heights, as at Poggio Garfagnana at about 400 metres (1300 feet) approximately half adopt the one device and half the other.

It will be noticed that we have here an instance, one out of many, in which it has been possible to calculate selection-pressures in nature. Two further examples of this may be given, differing from one another somewhat in type.

The first of these relates again to *Maniola jurtina*. The butterfly was studied by Dowdeswell and Ford on Tean, one of the smaller of the Isles of Scilly: it is about two and a half kilometres (half a mile) long and 16 hectares (40 acres) or so in area. After recording the spot-numbers for five years, we were fortunate to observe the effect of a major ecological change in the habitat. For in the autumn of 1951 a small herd of cattle, which had been maintained there for at least 150 years, was removed. Two areas of grass which they had kept grazed to an almost lawn-like condition then grew long, while regions of bracken and bramble which had been kept open by trampling, became an impenetrable jungle. The female spot frequency of the butterflies altered radically when this ecological change occurred. Since it had been constant annually at the old value and remained so at the new, the alteration evidently indicated an adjustment to the changed conditions. Its effect could be calculated, and demonstrated selection of about 60% against the non-spotted butterflies; evidently in response to the removal of the cattle for so long maintained on Tean.

We may here turn to material often cited as suitable for studies on ecological genetics at an elementary level; that is to say, the colour-pattern of the common snail *Cepaea nemoralis*. As will appear, this should in fact be used with considerable caution. The apparently simple facts of the situation are so generally known that they need only be summarized briefly in order to draw attention to a few features that tend to be insufficiently considered.

The ground-colour of the shell may be yellow, greenish when the animal is within; and this is recessive to the dark shades, pink or brown (Fig. 5–1). The genetic control is by multiple-alleles. Upon this background, dark bands may be added (Fig. 5–2). The bandless condition is dominant to that with any bands, the number of which (up to five) is determined by modifiers.

The two major loci thus responsible for shell-colour and banding constitute a super-gene, a situation discussed on pp. 12–13. The yellow shells are the less conspicuous to human eyes on a green background, the brown on a dark one, of fallen leaves for instance. Also, the bandless shells are the better concealed on a uniform background, such as downland grass, and the banded on a varied one: a mixed hedgerow, for instance. What is true for us is true also for an important predator of these snails, the Song Thrush, *Turdus philomelos*. Fortunately the birds carry the larger shells to convenient stones in order to break them open. There the remains accumulate so that it is possible to determine if the birds destroy the snails at random. This they do not do, showing that the specimens of *C. nemoralis* which are the less conspicuous to us are the less conspicuous to the birds also. One point here: only the larger shells are cracked open in this way for the smaller are swallowed whole. Thus the remains seen at 'thrush anvils' much underestimate the predation that takes place. Since the snails that do not match their background well are being differentially eliminated by the birds, the frequency of the less well adapted colour-patterns is being reduced in each locality. For instance, CAIN and SHEPPARD (1950) chose from among the many localities they studied the five with the most green at ground-level and the five with the most brown. Their results showed that:

The lowest percentage of yellow shells on a *green* background was 41.
The highest percentage of yellow shells on a *brown* background was 17.

Examining also the five most uniform habitats and the five most varied, they found that:

The lowest percentage of unbanded shells on a *uniform* background was 59.
The highest percentage of unbanded shells on a *variegated* background was 22.

But the elimination of the inappropriately coloured and patterned shells is taking place every summer. Why then do not the populations become

Fig. 5–1 *Cepaea nemoralis.* Yellow and brown forms on a dark background. (Slightly enlarged. Diameter at right angles to lip = 1.7 cm).

Fig. 5–2 *Cepaea nemoralis.* Yellow unbanded and yellow banded forms. (Enlarged as in Fig. 5–1.)

uniform instead of merely adjusting in frequency? The answer has already been provided in Chapter 2. For here we have a polymorphism, so that selective predation is only one of the agents which affects the survival of *C. nemoralis.* Heterozygous advantage must be another. Therefore there is a strong physiological bias in favour of mere diversity, even though it involves the survival of some of the less cryptic types.

Now there are large areas in England where the physiological effect completely outweighs the visual advantage. This is generally true on dry calcareous soils, such as chalk downs and overgrown sand dunes. It is true also over the greater part of France, which is hotter and drier than England. Banding may also be affected by climate. It has been found that the banded form increases at intermediate heights on the Pyrenees, where the climate is more equable than in the heat of the valleys yet milder than at higher and more exposed elevations. Thus it will be of much interest to find whether in any one locality colour-pattern or physiology is the more important. There are two further components which must be taken into account here.

First, many predominantly wooded localities pass from an average brown to an average green colour at ground-level as spring advances. Thus both colours may have an advantage in the same locality but at different times of year: a circumstance which tends to maintain diversity. It is one that can well be studied in practice, as was done most effectively by SHEPPARD (1951).

Secondly, as pointed out by de Ruiter in 1952, some birds build up a 'searching image'. That is to say, they hunt for prey resembling the one they last found, even to the exclusion of others that are the more obvious: a habit that promotes variability in common species subject to bird predation.

Here evidently in considering the snail we again, as in melanism (pp. 14–15, 19), encounter birds as selective agents. A further aspect of this now claims our attention. It is one on which new light has just been thrown by Dr. Miriam Rothschild.

Certain animals are 'aposematic'; that is to say, they tend to be protected from predators because they are distasteful, poisonous, or possess a defence mechanism such as a sting or barbed and irritating hairs. With these advantages, certain characteristics are associated. It evidently pays such forms to be easily recognized and so avoided, since they may be maimed or killed before their unpleasant or dangerous qualities are perceived. Thus, instead of being *cryptic*, or concealing, they have a *warning* colour-pattern, making them conspicuous. They move or fly slowly so as to give good opportunities for recognition; while, if insects, they may possess glands from which an offensive fluid will exude on attack. Moreover, as a very important adaptation, they are tough and leathery so that they can withstand crushing to a degree which would crack a brittle integument, like that of most butterflies, and cause death.

In general, females, especially of Lepidoptera, have evolved better protection than have males. For the female must often expose herself when egg-laying while she necessarily has the heavier body and so flies less swiftly and adroitly. Also in many species, one male can fertilize several females.

It is important to notice that these qualities can be adaptively adjusted so that they can be improved or minimized by selection acting on genetic variability. This can now be exemplified by the Scarlet Tiger Moth, *Panaxia dominula*, which is locally common in southern England (see p. 52). It flies by day, when its scarlet hind wings make it look almost tropical. That striking coloration is replaced by yellow in a rare variety, unifactorial and recessive. Rothschild (MARSH and ROTHSCHILD, 1974) has just demonstrated that the normal red females are at least three times as toxic as the yellow, this being the first time that a poisonous quality has actually been associated with a genetic difference. Unlike many species, the poison is not derived from the food (Comfrey, *Symphytum*) but manufactured by the insect itself.

It is to be noticed also that potential predators, birds and others, differ greatly in their susceptibility to toxins and their objection to a disagreeable taste; two features that are often but not always combined. Thus the American Blue Jay, *Cyanocitta cristata*, is so adversely affected by the larvae of the Monarch butterfly, *Danaus plexippus*, that having attempted to eat one it will starve to death rather than risk another experience of them. They contain the heart poisons calactin, which in addition to its toxicity has a bitter persistent taste, and callotropin. Conversely a single Quail, *Coturnix japonicus*, which is a small bird, will eat several of these larvae without apparent ill effects and is unharmed by a dose of digitalis sufficient to kill 50 men. It may be added that many birds have long memories and even after months will avoid eating an insect they have once tasted and disliked.

Unless, indeed, security is attained by an offensive smell, young birds must learn by experience what species are aposematic; and in spite of any resilience of their integument, that is an occurrence fraught with danger for the individuals they attack. The total destruction levied in that way can evidently be reduced if the aposematic forms resemble one another so closely that they are more or less indistinguishable. A single lesson of inedibility may then have a wide application. It is to be noticed in addition that though birds are close and accurate observers, an inexact similarity may also have some value when insects are seen on the wing.

Here we have the phenomenon of *mimicry*, which may extend across wide differences in classification. For all that is needed is, in general, visual resemblance. The red, for instance, of one group of Lepidoptera may be chemically very different from that of another: but that fact, revealing to a biologist, will not influence a potential predator. We are, indeed, beginning to realize that sound and scent may be important in mimicry, though little is known of those aspects of the subject so far.

We may first take some of Rothschild's remarkable new discoveries (Marsh and Rothschild, l.c.) and let them guide us in our understanding of mimicry since they relate to some of the best known British butterflies. This is particularly useful, as that phenomenon is to a great extent one of the tropics and sub-tropics, with their intense biological competition;

and though the condition throws such a clear light on the genetics of adaptation, it is not too easily demonstrated in temperate countries.

Mimicry can, then, be illustrated from the British White Butterflies, the Pieridae. The Large White, *Pieris brassicae* (Fig. 5–3a), is conspicuous as well as strongly toxic and unpleasant in all its stages, due largely to the mustard oils which it obtains from its food. This applies even to the eggs, which are yellow and laid in clusters. The larvae feed fully exposed on cabbage and are conspicuous also, as are the pupae. The white imagines are of course extremely obvious. When 250 mg of the female were ground in saline and injected intraperitoneally into mice, they caused death in 30 hours or so. About 342 mg of a similar male extract did so in four days.

The Small White, *P. rapae* (Fig. 5–3b), is less poisonous. The eggs are laid

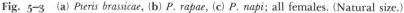

Fig. 5–3 (a) *Pieris brassicae*, (b) *P. rapae*, (c) *P. napi*; all females. (Natural size.)

singly and are colourless. The larvae are cryptic, for they are green and burrow into the cabbage plants or feed on the underside of the leaves; but there seems some doubt as to what extent they are toxic or distasteful. The pupae have a double protective mechanism, for they are toxic and an interperitoneal injection of 96 mg killed a mouse in 35 hours. Yet they also depend on concealment and actually have the capacity of adjusting their colour to match their background, whether green or brown, as illustrated by FORD (1971, Plate 2). However, the imagines had no lethal effect when an extract of the females, but only of 88 mg, was injected into a mouse. Yet as Rothschild says, and considering the quantity used in the *P. brassicae* injections, that result can only be accepted with great caution. We may take it in fact that had larger amounts of the imaginal extract been employed, toxicity would have become apparent; since poisonous mustard glycosides have actually been detected in the perfect insects of *P. rapae*. These have not indeed been found in any stage of the other Pieridae tested: including the Green-veined White, *P. napi* (Fig. 5–3c), and the Orange-tip, *Anthocharis cardamines*; nor, accordingly, have extracts from them had any detectable effect when injected into mice which, however, have some resistance to mustard oils.

We may now consider the general bearing of these results. The mimicry so far discussed is of the 'Müllerian' type, in which two or more protected species resemble one another in order, as already indicated, to reduce the effects of predation. There is also 'Batesian mimicry', when an aposematic form, which then becomes a *model*, is copied by an unprotected one, known as its *mimic*. The latter thus derives a benefit which it does not deserve. Rothschild points out that two of the adaptations successfully used, in general, by protected species cannot be employed in a true Batesian mimic. That is to say, a tough integument and an invariably slow and displaying flight; for their warning colour-pattern is a sham so that it cannot sustain close examination by a predator. Thus in the face of immediate danger, true Batesian mimics must be able to turn to rapid and cunning flight. However, the Müllerian and Batesian situations can overlap.

Let us relate these points to the British White Butterflies, Pieridae, so far mentioned. The larvae and pupae of the Large White are conspicuous, and this is the only one of the species in which the early stages are aposematic. For, as Rothschild points out, larvae and pupae, unlike imagines, have no means of escape from a predator. Consequently they need to be powerfully distasteful if they are to rely upon a warning colour-pattern for protection.

The imagines of the Small and Large Whites seem clearly to be Müllerian mimics, though depending considerably upon the greater distastefulness of the latter. The Green-veined White is much more of a Batesian mimic of the Large White; so too is the female Orange-tip; but not the male, with its striking orange patch. Indeed Batesian mimicry is generally absent from the male sex in butterflies, in part for the same reason as the sexual difference in aposematic qualities already mentioned (and see p. 4).

The Pieridae, therefore, tell us that there can be a seriation across from highly distasteful to very edible forms. This is further facilitated by the fact that a species which is strongly repellent to one predator may be acceptable to another. We are not, however, to think that the clearly aposematic way of life at one end of the series and the protective coloration fostered at the other, with or without mimicry of the strongly distasteful forms, are merely qualitatively different extremes in a range of toxicity and distastefulness. For they each have distinct qualities which relate to them alone, as aposematic or as cryptic species whose mimicry is respectively of the Müllerian or the Batesian type. For as Rothschild points out, the highly distasteful forms can, and do, develop a tough and leathery cuticle and slow flight, which allows them to be pecked, and their unpleasant body fluids to exude; and so repel, without serious damage to themselves, a predator which has actually attacked them. On the other hand, that form of adaptation is meaningless and very dangerous, and therefore not developed, at the other end of the scale, since the most cryptic species have little distastefulness on which to rely. Moreover, a Müllerian mimic will still retain its conspicuous appearance and aposematic habits if it extends its range beyond that of others within its mimicry complex. A Batesian mimic, on the other hand, will not in such circumstances continue to display an accurate mimicry but will become variable in the absence of its model, as will be exemplified on pp. 48–9. The truly Batesian mimic will, moreover, often be polymorphic, so as to prevent a given colour-pattern from becoming associated principally with edibility by a predator. Conversely, a Müllerian and powerfully protected species will not be polymorphic: it has eveything to gain from uniformity.

The fact that there may be a few birds or other forms which find a cryptic species unpalatable, or a normally toxic and strongly distasteful one edible, does not involve a breakdown in the two distinct mimicry situations. It is the average effect of predation in any district which will chiefly be important. We may now consider an example of polymorphic mimicry.

Papilio dardanus is a large butterfly, of 8 to 9 cm across the expanded wings. It is widespread in Africa south of the Sahara and continues up the coast to Sierra Leone, it is found also in Abyssinia. In addition, the species occurs in Madagascar, where both sexes are monomorphic, non-mimetic and very similar. On the upper surface they are yellow with black markings, while the hind wings are tailed.

This is true of males on the South African mainland, but not of the females. For they are polymorphic and tailless there. The various forms differ so much from one another and from the males that they were long thought to belong to different species. The females are generally mimetic, but the males never. Both sexes are tailed in Abyssinia.

Genetically the different female forms are controlled by a series of autosomal genes which give the impression of multiple alleles (p. 20); though they must in fact, occupy very closely linked loci and constitute a super-gene. Their action is sex-controlled to the females, in which dominance of the characters involved is usually though not always apparent. It is not possible to determine by inspection to which of these genotypes the males belong. The absence of tails in the female forms is due to the circumstance that their models are tailless. Indeed in the Oriental *Papilio memnon* the females are also polymorphic, and are respectively tailed and tailless according to the condition of the butterflies they copy.

In *P. dardanus* the females are very convincing mimics. Two of them (Fig. 5–4b and d) are shown with their models, and with the male (Fig. 5–4a). Their paler markings, on a black background, are brown in *trophonius* (b) while its minute spots and pale patches near the tips of the

Fig. 5–4 (a) *Papilio dardanus* male; top centre. The remainder numbered horizontally in two pairs. (b). *P. dardanus*, female form *trophonius*; a mimic of (c). *Danaus chrysippus*. (d). *P. dardanus*, female form *cenea*; a mimic of (e). *Amauris echeria*. (Reduced: actual size of **a**, wing tip to wing tip, =9.0 cm.)

fore-wings are white, so as to produce a good copy of *Danaus chrysippus* (c). *Trophonius* is dominant to *cenea* (d) which has white markings except for the patch near the base of the hind wings which is pale yellow. The model of *cenea* is *Amauris echeria* (e), together with the very similar *A. albomaculata*. Both these models belong to a highly distasteful family of butterflies, the Danaidae, while the *Papilio* is preyed upon by birds.

We find here in Batesian mimicry marvellous examples of adaptation. The condition must presumably be multifactorial and so attained gradually when monomorphic, as in the Sesiidae: moths which closely resemble wasps and other Hymenoptera. When Batesian mimicry is polymorphic it must, however, be under a switch-control supplied by a single gene or a super-gene; so as to convert one form to another without intermediates. Here we meet an evident difficulty. The adaptations required in order to resemble an unrelated model are bound to be complex. They often involve more than colour and pattern, extending to shape, as in *Papilio memnon* already mentioned, and to habit. A clear instance of the latter adjustment is provided by another African butterfly, *Hypolimnas dubius*. Its two forms copy the distinct flights of their models, respectively flapping and floating, and are the one shade-loving and the other sun-loving; yet on breeding, the two segregate on simple Mendelian lines.

In all these polymorphisms, the alleles which convert one phase into the other, or the units of the super-gene which does so, if this has been evolved, arise suddenly by mutation. Yet we cannot suppose that the mimics, with their complex attributes, do so too; for they would then have to depend upon the chance occurrence of a mutant evoking all their appropriate adjustments: an almost unimaginable occurrence even in a single instance. On the contrary, as pointed out originally by R. A. Fisher, once a gene chances to give some slight resemblance to an aposematic form, the similarity could be improved by selection acting upon the gene-complex (p. 22). Thus the mimicry evolves gradually, under selection, while remaining within the control of a single switch-unit.

CLARKE and SHEPPARD have tested this theory in a number of instances and fully confirmed it. We may take their work on *Papilio dardanus* (1960) as an example. It has already been mentioned that this species, with its multiple mimetic forms on the African mainland is yet monomorphic and non-mimetic in Madagascar. Clarke and Sheppard successfully mated mimetic females of known genetic constitution from South Africa with Madagascan males. The mimetic pattern breaks down and becomes confused in the progeny of that cross, since the gene responsible for it, whether *trophonius, cenea* or some other form, is then operating in a gene-complex which is not adjusted to it, being half Madagascan. On repeated back-crossing of such imperfect forms to South African material, the correct mimetic colour-pattern is gradually recovered. It cannot be doubted therefore that the genetic background of the South African race has been adjusted to improve its polymorphic mimicry; and that within the switch-mechanism controlling the different phases.

Since Müllerian mimicry favours uniformity so that any experience of distastefulness or toxicity shall have as wide an application as possible, it not only tends to produce close similarity between distinct aposematic species but opposes variability within each. Therefore Müllerian mimics will not normally be polymorphic.

The reverse is generally true, though in a rather specialized sense, of the Batesian situation. Evidently its existence threatens the security of its model and, furthermore, dilutes the advantage of copying it. For as the relative numbers of a Batesian mimic increase, its conspicuous warning colour-pattern comes to be associated more and more by predators with edibility than distastefulness. There will, in consequence, be a tendency in these circumstances towards polymorphism, so as to copy a number of distinct aposematic species. On the other hand, selection must favour sufficiently accurate resemblance to ensure that each mimetic form is mistaken for its model. There will therefore be a combination of diversity and constancy in Batesian mimicry. That is to say, its polymorphic phases will each be markedly invariable though differing greatly from one another.

We have seen in Chapter 2 that polymorphism involves heterozygous advantage, and of this Clarke and Sheppard have obtained clear evidence in mimicry. Just as the colour-patterns of snails may in some circumstances be immaterial compared with the physiological advantages of the heterozygotes (p. 41), so we find a corresponding situation in the polymorphism of butterflies. Non-mimetic species are not infrequently polymorphic, while not all the polymorphism of Batesian mimics may be mimetic. That statement is true of *Papilio dardanus*, for several of its phases have no models; for instance, *dionysos*, which occurs in West Africa.

The concept of mimicry, though now fully established, for long met with strong opposition. Those who rejected it pointed, among other things, to this very fact; saying that had *dionysos* chanced to resemble some aposematic butterfly, it would have been considered mimetic. There is a fallacy in that point of view for, in striking contrast with the mimetic polymorphic phases, the non-mimetic ones, including *dionysos*, are highly variable within themselves. So too, and this is a most telling fact, are the mimics if they extend their range beyond that of their models, surviving there by reason of heterozygous advantage alone.

Papilio dardanus occurs at high elevations in the mountains east of Lake Victoria where its models are rare or absent. The reality of its mimicry is demonstrated by the fact that there its mimetic forms, including *trophonius* and *cenea*, are extremely variable though so constant in the presence of their models. Thus a situation which when studied superficially might seem to contradict mimicry-theory in fact provides powerful evidence for it. It does so too for the evolution of heterozygous advantage, for the butterfly does not lose its polymorphism in the absence of its diverse models.

Another general feature of polymorphism is of course applicable when applied to mimicry, the tendency for its control to become supergenic, and there is evidence for this in *P. dardanus*. All its diverse polymorphic forms that have been studied genetically are controlled by what appears to be a series of multiple alleles, to the number of at least twelve: all of which give very much the effect of being situated at the same locus, that is to say. This would be a most improbable situation if true allelism were involved, but it is now clear that the loci concerned have merely been brought very close to one another. Indeed several of the polymorphic forms may best be interpreted as the result of rare cross-overs within such a super-gene.

Actually, the evidence obtained by Clarke and Sheppard for the supergenic control of mimicry is in some species remarkably complete; indeed in *Papilio memnon* they have even been able to determine the order of some of the included genes. One of these is that responsible for the presence of tails in the form which copies a tailed model. In this respect, these authors have provided a beautiful example of the genetic control of adaptation. In the island of Palawan there is an exceptional form of the butterfly which is always tailed. That condition is controlled by a different gene from the one responsible for the tailed mimic and does not have to segregate with a particular colour-pattern. It is of great interest to find therefore that, unlike the other gene for tails, it has not been brought into the super-gene responsible for mimicry.

It must be a matter of fundamental adaptation that Batesian, but not Müllerian, mimicry is so generally restricted to the female sex, never to the male; though there are a few instances in which both sexes are Batesian mimics. The point has already been mentioned on page 41, where two reasons for this situation are advanced. Two others may be mentioned here.

In the first place, the sexual stimulus in butterflies is partly olfactory and partly visual. There is therefore a reason to preserve a given male appearance: the one which has been evolved to stimulate the response of the female. Secondly, ROTHSCHILD (1971) points out that the limitation of mimicry to the females enables them to be twice as numerous relative to their models as they could be if both sexes were mimetic; and it is the females that are in greater need of protection.

It is noteworthy that in moths the colour-pattern plays little part in sexual attraction, which is mainly due to scent. Consequently male polymorphism is possible in these other superfamilies of the Lepidoptera, though it seems quite rare. A well-known instance of it is provided by the Wood Tiger, *Parasemia plantaginis* (Arctiidae). In that species the normal males have cream-coloured fore-wings while the hind pair are deep yellow, both having black markings. But there is an alternative male form, *hospita*, in which the ground-colour of the fore- and hind-wings is white. It is unifactorial and dominant, and the gene evidently has other effects, for this morph is strictly confined to arctic conditions and, further south, to mountains. In England the species is widespread on heaths, rough ground and openings in woods; but *hospita* is unknown except at a high elevation in the Lake District where, perhaps, 5 to 10% of the males are of that form. At that frequency, they will, effectively, all be heterozygotes, and it has been asserted that the homozygotes obtained in bred families are of poor viability. Either this is a mistake or, quite possibly, the effect of the gene is more favourably adjusted in some regions, for there is an area in Lapland where all the males are *hospita*. That form never occurs in the females. These are quite variable as to the extent of the black markings; while red pigment may be present on the body and sometimes in addition on the hind-wings. These diverse female forms are, however, mainly local and are not, at any rate, polymorphic.

Unisexual polymorphism, whether restricted to the male or the female, is not connected with the XX, XY chromosome mechanism but is autosomal and sex-limited. Thus the expression of the controlling gene is restricted to a particular internal environment, that provided by one of the sexes. This need not cause surprise; for we are familiar with just that situation in the accessory sexual characters of the male and the female in general.

On reviewing what has been said on polymorphism so far, certain aspects of that condition will have become apparent, and these may now be analysed in the light of additional examples. The super-gene may first be considered in this way.

This device has been mentioned in *Primula* and in butterfly mimicry. In some other instances its existence is uncertain, as in *P. plantaginis*. We simply do not know whether the qualities which restrict *hospita* to high latitudes and altitudes are due to selection involving the multiple effect of the controlling gene or, more probably, to other genes segregating with that for male coloration, these being held together supergenically.

It is obvious enough that with the great power of selection which now faces us, so different from what had been contemplated in the earlier part of this century, we may expect to find the evolution of the super-gene in some species and yet to meet others closely related in which this arrangement has not yet been attained. Though in some of these it would appear to be of considerable advantage.

Sphaeroma, a genus of marine Crustacea (Isopoda), reaches its northern limit along the south and west coasts of England. Consequently, we find, as might be expected (pp. 25–6), striking distinctions in colour-pattern even in adjacent localities. Many of these variants are polymorphic. The species which has undergone the least adjustment to that situation is *S. serratum*, in which the loci controlling the different morphs are unlinked. However, in others related to it, *S. mondi, S. bocqueti* and *S. rugicauda*, for instance, the controlling genes are enclosed within a super-gene, and rare cross-overs have been detected within it. It is worth noting that two of the colour-patterns, known as 'ornatum' and 'signatum', which are unlinked in *S. serratum*, seem to fall within a super-gene in *S. bocqueti*. Yet a more searching examination has shown that in the latter species 'ornatum' is produced by another allele with a similar phenotype; we are reminded here of the tailed condition of *Papilio memnon* in the island of Palawan when compared with its occurrence elsewhere (p. 49).

Certain of the *Sphaeroma* genes are clearly adjusted to the environment. Thus that for yellow coloration in *S. rugicauda*, lethal when homozygous, gives better survival at low temperatures than does its allele for a grey body-colour.

The *Sphaeroma* species are easy to breed though rather slow, taking about seven months to reach maturity, nor are they difficult to find. They inhabit salt marshes, are often common, and can be caught by means of a hand net, from areas of turf when covered at high tide. Detailed information on their polymorphism may be obtained from the references given here, and interesting work on the inter-relationship of the forms and their survival in varied external conditions could easily be carried out.

The advantages, then, of a super-gene enclosing the loci concerned with body-colour have evidently been made use of in most of the *Sphaeroma* species, but not in *S. serratum*. One might suspect that this would be at a considerable handicap without that device. Yet the disadvantage has to some extent been overcome for, as Sheppard has pointed out, its necessary combinations of characters are attained by the alternative though less satisfactory method of interaction between the different polymorphic forms.

It will have been noticed that polymorphism affecting directly observable characters, such as colour and pattern, may also control habit, as in melanism and mimicry. This is evidently an important aspect of adaptation. It has been studied in detail by Hovanitz in the alternative phases of certain Sulphur Butterflies of the genus *Colias*. These fall into two groups as to their main ground-colour; being either pale greenish to whitish in both sexes, or deep orange. In the latter species, of which one, *C. croceus*, reaches southern England at intervals and occasionally becomes common, the males are monomorphic. The females, however, in addition to the male-like orange specimens always have a less common alternative form that varies from a primrose shade to white. This is dominant in all the species, and is due to an autosomal gene sex-controlled in its action. It is quite possible that these pale females gain some protection from their resemblance to the White Butterflies, Pieridae.

Working with a North American species, *Colias eurytheme*, much resembling *C. croceus*, Hovanitz found that the gene controlling the dimorphic colouring affects also the habits of the imagines: the whitish females are relatively more active than the orange in the early morning and in the evening. They are also relatively commoner at high altitudes and latitudes, and they tend to emerge from the pupa in advance of the normal form.

There is probably a complex ecological balance here. We do not know if the habit difference is operative also in the males as the two genotypes cannot be distinguished in that sex.

Protein variation, in which normally invisible polymorphisms can be detected by means of electrophoresis (p. 36) is, as already indicated, likely to disclose a situation of value for genetic analysis. Indeed we already know that though cryptic in its effects it may be involved with other characters of selective importance, including habit. Thus protein diversity in the Eiderduck, *Somateria mollissima*, is related to the migration of that bird.

Numerous instances have now been given in which polymorphism is adjusted in various ways within the switch-control of a pair of alleles or of the phases of a super-gene. For selection can operate on variation arising from segregation within the genetic background of the organism, and so modify the expression of distinct polymorphic forms without affecting their occurrence as clear-cut alternatives. Some of the earlier experimental work of this kind to be conducted in the laboratory was carried out on the Scarlet Tiger Moth, *Panaxia dominula*, already mentioned (p. 42). The species is widepread in southern England as localized colonies chiefly along the banks of rivers or in marshes. It is only known to be polymorphic in a single locality, at Cothill five miles from Oxford. Three forms exist there owing to the segregation of a pair of alleles without full dominance. The heterozygotes, known as *medionigra*, much resemble the normal form but are distinguishable from it owing the reduction or absence of the central cream-coloured spot on the fore-wings and to the presence of an extra black dot on the hind pair. It generally occupies only a small percentage of the population, varying from year to year, so that the homozygotes are of great rarity. They are very abnormal in appearance owing to a heavy excess of black pigment.

The species has one generation in the year. It was possible by means of artificial selection for four generations to increase the characteristics of the heterozygotes in one line and to diminish them in another. That is to say, to make the effect of the gene more dominant or more recessive. Crosses between the two selected strains and normal individuals from the colony recovered the original appearance of *medionigra* and so showed that the switch-gene itself had not been altered but rather the genetic background in which it worked.

The special interest of this study arises from the fact that a comparable change towards dominance was taking place in the original wild population and towards recessiveness in a neighbouring, but isolated colony where the *medionigra* form had been introduced. This indeed seems to be the only instance in which an evolutionary adjustment taking place in nature has been forestalled experimentally.

6 Genetic Adaptations in Man

It is appropriate to close this brief account of genetics and adaptation with a few references to the human situation. A subject which may be further illustrated by doing so is one to which repeated reference has been made, and was indeed discussed in the first chapter of this book: the significance of selection compared with mutation and random survival.

Elementary accounts of genetics often use colour-blindness to introduce the idea of sex-linkage; that is to say, the situation in which genes are carried in the non-pairing region of the X-chromosome and are therefore related in their segregation to sex.

This abormality is remarkably common, for in north-western Europe and other civilized areas about 8% of men is colour-blind. The condition is recessive and is consequently much the more frequent in the heterogametic sex, that with a single X-chromosome. This is the male in mammals though the female in birds. For in the homogametic sex, XX, the defect will generally be obscured by normal colour-perception since that is dominant to it. Assuming equal viability of the two genotypes, a recessive sex-linked condition is distributed in men and women as $p : p^2$; so that, in this instance, one colour-blind individual is to be expected in every hundred and sixty-three women.

The situation is in reality more complex, since there are two genes for recessive colour-blindness, one affecting the ability to distinguish green and the other to distinguish red. They are not alleles but are carried at different loci in the X-chromosome. The already mentioned frequency of 8% of men refers to the two types together; of these, green colour-blindness is in fact the commoner. (There are also two forms of each: one more and one less extreme, due to multiple alleles at both loci.)

Colour-blind individuals prove to be rarer in primitive races. Thus only about 2% of male Australian aborigines, Fijians and North American Indians are affected in this way. POST (1962), who drew attention to that situation, reasonably suggests that this lower value is due to selection against colour-blind men, who will in general be at a disadvantage among those who depend upon hunting and food-gathering. He, however, considered that the normal higher frequency indicates selective neutrality, the condition being of no importance in a civilized community. From what has already been said, it will be realized that this is a quite impossible assumption and that a selectively neutral quality could not have spread in this way. On the contrary, it will be held in equilibrium balanced at the higher level also; by heterozygous advantage in women and, furthermore, by other effects of the gene. Even these may operate in a balanced way; a fact suggested indeed by a curious circumstance. That is to say, those affected by colour-blindness can appreciate certain colour-distinctions imperceptible to the ordinary individual. It is indeed revealing to go into the country with a trained but colour-blind naturalist and to note the extraordinary ability to detect cryptic animals, for instance Lepidoptera, which he will occasionally display. This, however, is an attribute of limited application, and the colour-blind are in general at a handicap in regard to colour-perception.

We can now turn to an interesting but complex subject: that of the human blood groups. Though such serology is in the main much beyond the standard to be covered in this book, a few words on its adaptive significance are highly appropriate here, especially as the subject is so widely discussed today.

It is necessary to realize that there are thirteen or more blood group series in Man, each containing a number of interrelated blood groups. These are controlled by major genes, sometimes by super-genes, as polymorphisms, with all that this implies. Therefore the various groups within each series are balanced by contending advantages and disadvantages. What are they?

In the first place, the heterozygotes must, on theoretical grounds, be at an advantage compared with the homozygotes: a fact for which decisive evidence has repeatedly been obtained, in spite of the dominance which tends to mask that situation. Further, there are powerful interactions between the groups. Thus the blood of a mother can kill her child or that of a child can kill its mother: for though the blood circulations of mother and embryo are separate, slight placental leaks occur, which are partly responsible for selective elimination against incompatible foetuses (those whose blood group interacts dangerously with that of the mother). Yet, on the other hand, combinations of certain of the blood groups are advantageous.

Furthermore, there is a tendency for blood groups to be associated with liability to develop specific diseases. We may take two instances out of many detected in the best known series, OAB. Those who belong to group O (comprising about 43.5% of the population in southern England) are more liable, by about 30%, to duodenal ulcers than are those of the other types (groups A, B and AB). A comparable fact, but recently established, must have had a profound effect upon the evolution of the human races. That is to say, those who belong to groups A and AB (their frequencies are, respectively, 44.7 and 3.2% in southern England) are much more liable to develop smallpox, and to do so in a more severe form, than those of groups O and B: this chiefly during epidemics in unvaccinated populations.

We can here turn to an entirely different aspect of human blood. Haemoglobin, the red oxygen-carrying pigment, exists in several types. There is a foetal form, which begins to be replaced by the adult one shortly before birth. It has normally disappeared completely after a year. Adult haemoglobin consists of a major part A1 and a minor A2. The A1 fraction can be substituted by an abnormal 'sickle-celled' alternative, controlled by a pair of alleles. When homozygous, the red corpuscles depart from their ordinary circular outline, being long and curved with slender thread-like processes. These homozygotes die of anaemia during childhood or youth. In the heterozygotes, on the other hand, the red corpuscles look normal in circulation but they become sickle-shaped if air be excluded from a drop of the blood. Thus all three genotypes can be recognized.

In Greece, the heterozygotes amount to about 17% of the population, and in some African tribes up to 40%. This is extraordinary, considering that the condition is fatal in the homozygotes: a handicap which must evidently be balanced by some powerful advantage. Its nature is known, for the heterozygotes are to a considerable degree protected from malaria in its most dangerous form (that due to *Plasmodium falciparum*). It is only where that disease is prevalent that sickle-celled anaemia is found.

Populations have now been discovered in Saudi Arabia possessing the same gene for the sickle-celled condition but in which even the homozygotes have become fully viable. The ordinary heterozygotes in Africa carry a trace, variable in amount, of foetal haemoglobin which they ought to lose in infancy. Selection has operated on its variability to provide these Arabs with enough, 18% on the average, of the foetal type for nearly normal life; for this seems almost equal to the A1 form in quality, while sufficient sickle-cell haemoglobin persists for protection against the malaria.

Here we encounter something impressive: a human adaptation seen in its evolving stages. In Africa the homozygotes die that the heterozygotes may live, by gaining the sickle-cell advantage. Yet the ingredients are there, the trace of foetal haemoglobin and its variability, no doubt polygenically determined, to produce the perfect article that we find in Arabia: the benign sickle-celled condition, protected against malaria yet preserved from homozygous lethality.

An instance of adaptation such as that just described leads us to the fundamental thought of human evolution itself. What have been the basic requirements for it? Of course a number could be cited, but a few of them are outstanding. One, at the chromosome level, has already been mentioned (p. 27), and others of a similar type exist. For instance, the evolution of intensely fluorescent chromatin separates man, the chimpanzee and the gorilla from all other Hominids. That, however, is evidently not a human attribute alone. On the other hand, of the twenty-three pairs of human chromosomes, five (identified as 2, 4, 9, 17 and 18, numbered in sequence according to size) are present only in man or in man and the chimpanzee; excluding, that is to say, the gorilla.

We can, however, profitably consider what features have been the prerequisite of human evolution, singling out man as a creature leading, above all, a life of thought. Manifestly, that depends upon slowing down development and so extending youth and the period of learning: at the age of a few weeks a puppy is a far more advanced creature than is a human child.

I once had the privilege of dissecting several thousand wild mice, *Apodemus sylvaticus*, and examining, therefore, large numbers of the females in every stage of pregnancy. One feature struck me, as it would anyone else; that early in the process, the uterus contained not only more embryos than were ever born but more than possibly could have reached term, owing to the space available. Later, one saw that the embryos were growing at different speeds. It is true that there may be some slight advantage in uterine position, depending on how near to a suitable blood vessel an implant may be; but on the whole the environment of the mammalian uterus is a singularly constant one. Most of such variation must therefore be genetic. Moreover, at the date of which I write, the discovery had just been made that genes can control the time of onset and rate of development of processes in the body (FORD and HUXLEY, 1927). Thus the existence of such *genetic* variation had then recently become apparent; when affecting the organism as a whole, it would doubtless be polygenic. On turning to later stages in pregnancy, one saw the larger embryos trespassing on the blood supply of the smaller and, later still, the latter being resorbed and destroyed.

Evidently in such an animal as this mouse rapid development must carry all before it. However excellent the genotype of a particular embryo may be, there can be no opportunity of exploiting it if occurring in an individual that is a relatively slow developer. Thus the possibility of extending the youthful phase, basic as that must be to the evolution of Man, could never have arisen if a premium had been put on rapid development in the Hominid line, as it would have been in the face of embryonic competition.

It has, then, been essential that the human race and its immediate ancestors should have produced, on the whole, but one embryo at a conception; so freeing it from competition and allowing it a relatively long intra-uterine period. The effect of genes which promote the latter situation would tend to continue subsequently, when it could be reinforced by others of similar tendency but operating principally after birth.

Here indeed are fundamental aspects of human evolution. They depend upon the fact that both the number of eggs discharged from the ovaries at a given time, and rates of development in the body, are under genetic control. Thus it has been possible for selection to lengthen the baby phase, youth and the period of learning; and so give rise in a somewhat primitive, and therefore generalized, type of mammal to new and transcendent qualities.

References

ANTONOVICS, J. and BRADSHAW, A. D. (1970). Clinal patterns at a mine boundary, *Heredity*, **23**, 219–38.

BIRCH, L. C. (1955). Selection in *Drosophila pseudoobscura* in relation to crowding. *Evolution*, **9**, 389–99.

BRADSHAW, A. D., MCNEILLY, T. S. and GREGORY, R. P. (1965). Industrialisation, evolution and development of heavy metal tolerance in plants. *5th Symp. Brit. Ecol. Soc.*, 327–43, Blackwell, Oxford.

CAIN, A. J. and SHEPPARD, P. M. (1950). Selection in the polymorphic land snail *Cepaea nemoralis*. *Heredity*, **4**, 275–94.

CLARKE, C. A. (1964, 2nd edn.). *Genetics for the Clinician*. Blackwell, Oxford.

CLARKE, C. A. and SHEPPARD, P. M. (1960). The genetics of *Papilio dardanus*. *Genetics*, **45**, 439–57.

CLARKE, C. A. and SHEPPARD, P. M. (1966). A local survey of the distribution of industrial melanic forms in the moth *Biston betularia*. *Proc. roy. Soc.*, B, **165**, 424–39.

CREED, E. R., (1971). Melanism in the Two-spot Ladybird, *Adalia bipunctata*. In *Ecological Genetics and Evolution* (ed. E. R. Creed), pp. 134–51. Blackwell, Oxford.

CREED, E. R.; LEES, D. R. and DUCKETT, J. G.(1973). Biological method of estimating smoke and sulphur dioxide pollution, *Nature, Lond.*, **244**, 278–80.

CROSBY, J. L. (1940). High proportions of homostyle plants in populations of *Primula vulgaris, Nature, Lond.*, **145**, 672–3.

DARLINGTON, C. D. (1963, 2nd edn.). *Chromosome Botany*. Allen & Unwin, London.

DARLINGTON, C. D. and MATHER, K. (1949). *The Elements of Genetics*. Allen & Unwin, London.

DARWIN, C. R. (1877). *The different forms of flowers on plants of the same species*, Murray, London.

DOBZHANSKY, TH. (1970). *Genetics of the Evolutionary Process*. Columbia Univ. Press, N.Y.

FALCONER, D. S. (1960). *Introduction to Quantitative Genetics*. Oliver & Boyd, Edinburgh.

FISHER, R. A. (1930). The distribution of gene ratios for rare mutations. *Proc. roy. Soc. Edinb.*, **50**, 204–19.

FORD, E. B. (1940a). Genetic research in the Lepidoptera. *Ann. Eugen.*, **10**, 227–52.

FORD, E. B. (1940b). Polymorphism and taxonomy, pp. 493–513, *The New Systematics* (ed. Julian Huxley), Clarendon Press, Oxford.

FORD, E. B. (1971, 3rd edn. reprint). *Butterflies*. New Naturalist Series, Collins, London.

FORD, E. B. (1973, 7th edn.). *Genetics for Medical Students*. Chapman & Hall, London.

FORD, E. B. (1975, 4th edn.). *Ecological Genetics*. Chapman & Hall, London.

FORD, E. B. and HUXLEY, J. S. (1927). Mendelian genes and rates of development in *Gammarus chevreuxi*. *Brit. J. Exp. Biol.*, **5**, 112–34.

GORDON, C. (1935). An experiment on a released population of *D. melanogaster*. *Am. Nat.*, **69**, 381.

GRAIN, D. (1958). Radiation response in mice. *Genetics*, **43**, 835–43.

GREENBERG, J. (1964). A locus for radiation resistance in *Escherichia coli*. *Genetics*, **49**, 771–8.

GREGORY, R. P. and BRADSHAW, A. D. (1965). Heavy metal tolerance in populations of *Agrostis tenuis* and other grasses. *New Phytol.*, **64**, 131–43.

HANDFORD, P. T. (1973). Patterns of variation in a number of genetic systems in *Maniola jurtina*: the boundary region. *Proc. roy. Soc.*, B, **183**, 265–84.

IVES, P. T. (1950). The importance of mutation rate genes in evolution. *Evolution*, **4**, 236–52.

JACOBS, P. A., PRICE, W. H., RICHMOND, S. and RATCLIFF, R. A. (1971). Chromosome surveys in penal institutions and approved schools. *J. Med. Genet.*, **8**, 49–58.

KETTLEWELL, H. B. D. (1973). *The Evolution of Melanism*. Clarendon Press, Oxford.

KIMURA, M. (1968). Evolutionary rate at the molecular level. *Nature*, **217**, 624–6.

LEE, B. T. and HOLLOWAY, W. B. (1965). The genetic control of radiosensitivity in *Pseudomonas aeriguosa*. *Radiation Res.*, **25**, 68–77.

LEES, D. R. (1971). The distribution of melanism in the Pale Brindled Beauty Moth, *Phigalia pedaria*, in Great Britain, In *Ecological Genetics and Evolution* (ed. E. R. Creed), pp. 152–74, Blackwell, Oxford.

LEES, D. R., CREED, E. R. and DUCKETT, J. G. (1973). Atmospheric pollution and industrial melanism. *Heredity*, **30**, 227–32.

MARSH, N. and ROTHSCHILD, THE HON. M. (1974). Aposematic and cryptic Lepidoptera tested on the Mouse. *J. Zool. Lond.*, **174**, 89–122.

PARSONS, P. A., MACBEAN, I. T. and LEE, B. T. (1969). Polymorphism in natural populations for genes controlling radioresistance in *Drosophila*. *Genetics*, **61**, 211–18.

POST, R. H. (1962). Population differences in red and green color vision deficiency. *Social Biology, (Eugen. Quart)*, **9**, 131–46.

ROTHSCHILD, THE HON. M. (1971). Speculations about mimicry with Henry Ford. In *Ecological Genetics and Evolution* (ed. E. R. Creed), pp. 202–23, Blackwell, Oxford.

RUITER, L. DE (1952). Some experiments on the camouflage of stick caterpillars. *Behaviour*, **4**, 222–32.

SCALI, V. (1971). Spot-distribution in *Maniola jurtina* (Lepidoptera, Satyridae): Tuscan Mainland, 1967–9. *Monitore Zool. Ital.*, (n.s.), **5**, 147–63.

SHEPPARD, P. M. (1951). Fluctuations in the selective value of certain phenotypes in the polymorphic land snail *Cepaea nemoralis*. *Heredity*, **5**, 125–34.

SMITH, R. A. and BRADSHAW, A. D. (1970). Reclamation of toxic metalliferous wastes using tolerant populations of grass. *Nature*, **227**, 376–7.

WAHRMAN, J., GOITEIN, R. and NEVO, E. (1969). Mole Rat *Spalax*: Evolutionary significance of chromosome variation. *Science*, **164**, 82–4.

WESTERMAN, J. M. and PARSONS, P. A. (1973). Variations in genetic architecture at different doses of γ-radiation measured by longevity in *Drosophila melanogaster*. *(Can. J.). Genet. Cytol.* **15**, 289–98).